Metal Oxide Thin Films: Synthesis, Characterization and Applications

About the Editors

Erwan Rauwel received his PhD degree from the University of Caen in materials science in 2003. He continued with postdoctoral studies in collaboration with STMicroelectronics at Minatec, Grenoble, France and at the University of Aveiro (Marie Curie IE Fellowship), Portugal. He worked as a senior researcher at the university in collaboration with Statoil and then worked at Taltech as a professor in Estonia for 5 years. He is now a professor at the Institute of Technology of the Estonian University of Life Science in Estonia where his team investigates the properties of nanoparticles for water purification, the development of biocidal coating to fight AMR and COVID-19, and the development of nanocomposites for photocurrent generation and energy harvesting and they also investigate new possible treatments against cancer with the LBN (Montpellier). He has more than 65 peer-reviewed publications with an H-index of 22, 3 books, 5 book chapters, and 6 patents. He is chief scientist in his start-up company specializing in nanomaterials (PRO-1 NANOSolutions). More information about his recent research activities can be found at www.rauwel.eu.

Protima Rauwel is Head of Energy Applications Engineering Chair at the Institute of Technology of the Estonian University of Life Sciences. She obtained her PhD in 2005 from the University of Caen, France in condensed matter and materials science with a special focus on nitride thin films with quantum dots and quantum wells. She has postdoc experience from the University of Aveiro, Portugal and the University of Oslo, Norway, where she continued her research on thin films and freestanding nanomaterials. Her research focus is presently on nanomaterials for water remediation, optoelectronic applications, and anti-microbial applications with a focus on synthesis, characterization, and application. She has co-authored more than 85 publications, 3 books, and 5 book chapters. She is an inventor on 2 patents and has a Google Scholar H-index of 28.

 materials

Editorial

Editorial for the Special Issue on 'Metal Oxide Thin Films: Synthesis, Characterization, and Applications'

Erwan Rauwel * and Protima Rauwel

Institute of Technology, Estonian University of Life Sciences, Kreutzwaldi 56/1, 51014 Tartu, Estonia; protima.rauwel@emu.ee
* Correspondence: erwan.rauwel@emu.ee

Citation: Rauwel, E.; Rauwel, P. Editorial for the Special Issue on 'Metal Oxide Thin Films: Synthesis, Characterization, and Applications'. *Materials* **2021**, *14*, 1834. https://doi.org/10.3390/ma14081834

Received: 31 March 2021
Accepted: 6 April 2021
Published: 7 April 2021

Publisher's Note: MDPI stays neutral with regard to jurisdictional claims in published maps and institutional affiliations.

The last two decades have witnessed the development of new technologies for thin-film deposition and coating. A marked improvement in thin-film deposition methods, enabling the precise coating of surfaces at the nanoscale on complex nanostructure and on large surfaces is viable. These include new deposition methods like atomic layer deposition (ALD) that enables the coating on complex nanostructures with high conformality and homogeneity. Other more conventional methods based on coating in solution have also been improved in order to achieve a high quality conformal coating. These deposition processes also enable the synthesis of multilayers and the development of synergistic effects through the mutual interaction of such layers. At the same time, a need for new tools to characterize these films has emerged. This Special Issue is dedicated to these recent developments and contains 10 manuscripts reporting thin-film deposition and coating and their characterization. It compiles works on vapor-phase deposition processes including atomic-layer deposition (ALD) [1,2], mist-chemical-vapor deposition [3] and atmospheric-pressure chemical-vapor deposition (APCVD) [4] methods in which authors have studied the stabilization of metastable phases and the effect of growth conditions on thin film quality, epitaxy, and crystallinity. The Special Issue also includes liquid-phase coatings like sol-gel [5], solution immersion [6], micro-arc oxidation [7], Plasma Electrolytic Oxidation Processing [8], and single pulse anodization [9] where the authors were more focused on the coating of aluminum. The last paper reports on a new metal oxide thin film characterization method [10] that can extract the majority charge carriers' type, their concentration, and mobility and is extendable to the study of other wide-bandgap semiconductors. Two of these manuscript reports studies are on Transparent Conductive Oxide thin films [4,10] and in one manuscript, the authors investigate InGaZnOx Thin Film Transistors [6].

The first four manuscripts of this Special Issue consist of thin-film depositions in gas phase or solution. The stabilization of $NiTiO_3$ metastable phase using atomic layer deposition (ALD) was studied by Bratvold et al., in the manuscript "Phase and Orientation Control of $NiTiO_3$ Thin Films" [1]. They demonstrated that it is possible to stabilize $NiTiO_3$ disordered phase (R−3c) on α-Al_2O_3 substrates. They also investigated the influence of substrate on the symmetry and orientation of $NiTiO_3$ thin films under epitaxial growth conditions on $LaAlO_3(100)$, $SrTiO_3(100)$ and $MgO(100)$ substrates. They have demonstrated that the orientation and symmetry of $NiTiO_3$ thin films deposited by ALD can be controlled by the choice of a suitable substrate. Using ALD, Duparc et al., studied the deposition of $GdCoO_3$ and $Gd_{0.9}Ca_{0.1}CoO_3$ thin films on $LaAlO_3(100)$ and $YAlO_3(100/010)$ substrates in "Atomic Layer Deposition of $GdCoO_3$ and $Gd_{0.9}Ca_{0.1}CoO_3$" [2]. These coatings may find applications in the field of oxidation catalysis. They showed that crystal orientations of the films can be tuned by the selection of the substrate and mitigated through the interface via solid-face epitaxy upon annealing. In the third manuscript, Xu et al. report on the influence of carrier gases in mist chemical vapor deposition method on the quality of epitaxial corundum-structured α-Ga_2O_3 films [3]. The thin films were grown on c-plane sapphire substrates with single (0006) plane orientation for all carrier gases, but the authors demonstrated that the O_2 gas enables the growth of better quality epitaxial films with increased growth rate (10.3 nm/min vs. 5.3 nm/min and 2.4 nm/min) and better crystalline quality. The fourth manuscript reports the study of the influence of La doping in $BiFeO_3$ thin films grown on $Pt/TiO_2/SiO_2/Si$ substrates by the sol-gel method [5]. The authors have shown that it is possible to increase the grain size and self-polarization and promote (102) texture of the film by La doping and a higher growth temperature. These characteristic are important for future piezoelectric applications. They show that piezoelectric properties are connected to the films' growth conditions and La doping, emphasizing that if both are thoroughly controlled $BiFeO_3$ -based films could be used in micromechanical applications.

Manuscripts reporting on thin films grown by APCVD and solution immersion follow these studies. In "Origin of Magnetotransport Properties in APCVD Deposited Tin Oxide Thin Films" by Juraić et al., the structural and magnetotransport studies of tin oxide thin films prepared by the cost-effective Atmospheric Pressure Chemical Vapor Deposition method deposited on soda-lime glass substrates [4]. In particular, they studied the charge carrier density and mobility of the films that are typical values of SnO_2 films manifesting a temperature dependence in their case. This behavior is attributed to grain boundary scattering and crystallites preferred orientation. In the second paper, "Optimizing the Properties of InGaZnOx Thin Film Transistors by Adjusting the Adsorbed Degree of Cs^+ Ions", Zhang et al. investigated surface structure modifications and increased oxygen vacancy concentrations of a-IGZO thin films through the introduction of Cs^+ ions into the surface [6]. They observed an improvement of transfer properties and stability of the films coated by solution immersion method at low temperature. They found that the optimized performance of Cs-doped InGaZnOx Thin Film Transistors was obtained for a Cs^+ ion concentration of 2% mol/L, with thin films that exhibit a carrier mobility of 18.7 $cm^2 \cdot V^{-1} \cdot s^{-1}$, OFF current of 0.8×10^{-10} A and a threshold voltage of 0.2 V.

The last set of manuscripts focus more particularly on aluminum coating using liquid phase coating. In these investigations, the authors mainly investigated how the coating methods could be improved and identified the main parameters that control the growth quality. Song et al. studied the tribological performances of micro-arc oxidation of aluminum 6061 alloy with the addition of 0.75 g/L of cellulose into the electrolyte [7]. They observed an improvement of the coating with a decrease of the roughness along with an increase of cellulose concentration in the electrolyte. They indicate that cellulose fills the micro-cracks and micro-pores and a part of cellulose is cross-linked with the Al^{3+}. This study is followed by the report of Yang et al., in which the authors investigated a novel self-adaptive control method for large scale coating of aluminum alloys using plasma electrolytic oxidation processing [8]. They showed that coating can be controlled via feature information (Vb, Ib, Vi, and Vu) and the growth process can be tuned via the voltage, current, and processing time parameters. The coating exhibited good uniformity and compactness without visible cracks and large pores. For optimal growth process conditions on 6061 aluminum alloy samples, they measured a specific energy consumption of 1.8 $kWh \cdot m^{-2} \cdot \mu m^{-1}$. The same group also studied the discharge behavior and dielectric breakdown of oxide films deposited on aluminum in an alkaline silicate electrolyte using single-pulse anodization of aluminum micro-electrodes [9]. They measured voltage and current waveforms of applied pulses and studied the morphology of the coating. They found a good correlation between the pulse parameters and shape of discharge channels and showed that by increasing potential and pulse width it is possible to form circular opened pores on the surface.

The last paper entitled "Hot-Probe Characterization of Transparent Conductive Thin Films" describes the dynamic hot-probe measurement system that replaces or supplements existing methods like Hall effect measurements or the Haynes–Shockley experiments [10]. The replacement of indium–tin oxide by more cost-effective materials like ZnO doped by Al, Mn, or Sb requires mobile and low-cost evaluation methods. The paper shows that Dynamic Hot-Probe Characterization can extract the main parameters of metal oxide thin films like the majority charge carriers type, their concentration, and mobility and this method may also be applied to other wide-bandgap semiconductors.

We hope that this special issue will serve as a valuable support for researchers in the field of metal oxide thin-film deposition and characterization and we wish you a pleasant read.

Funding: This research was funded by European Regional Development Fund grant number "Emerging orders in quantum and nanomaterials" and Eesti Maaülikool (EMÜ) Bridge Funding grant number P200030TIBT.

Institutional Review Board Statement: Not applicable.

Informed Consent Statement: Not applicable.

Data Availability Statement: Not applicable.

Acknowledgments: The guest editors would like to thank the authors for submitting their work to the special issue and its successful completion. A special thank you to all the reviewers participating in the peer-review of the manuscripts submitted. We are also grateful to Tiffany Wan and the editorial assistants who made the whole special issue guest editing a smooth and efficient process.

Conflicts of Interest: The authors declare no conflict of interest.

References

1. Bratvold, J.E.; Fjellvåg, H.; Nilsen, O. Phase and Orientation Control of $NiTiO_3$ Thin Films. *Materials* **2020**, *13*, 112. [CrossRef] [PubMed]
2. Duparc, M.; Sønsteby, H.H.; Nilsen, O.; Sjåstad, A.O.; Fjellvåg, H. Atomic Layer Deposition of $GdCoO_3$ and $Gd_{0.9}Ca_{0.1}CoO_3$. *Materials* **2020**, *13*, 24. [CrossRef] [PubMed]
3. Xu, Y.; Zhang, C.; Cheng, Y.; Li, Z.; Cheng, Y.n.; Feng, Q.; Chen, D.; Zhang, J.; Hao, Y. Influence of Carrier Gases on the Quality of Epitaxial Corundum-Structured α-Ga_2O_3 Films Grown by Mist Chemical Vapor Deposition Method. *Materials* **2019**, *12*, 3670. [CrossRef] [PubMed]
4. Juraić, K.; Gracin, D.; Čulo, M.; Rapljenović, Ž.; Plaisier, J.R.; Hodzic, A.; Siketić, Z.; Pavić, L.; Bohač, M. Origin of Mangetotransport Properties in APCVD Deposited Tin Oxide Thin Films. *Materials* **2020**, *13*, 5182. [CrossRef] [PubMed]
5. Semchenko, A.V.; Sidsky, V.V.; Bdikin, I.; Gaishun, V.E.; Kopyl, S.; Kovalenko, D.L.; Pakhomov, O.; Khakhomov, S.A.; Kholkin, A.L. Nanoscale Piezoelectric Properties and Phase Separation in Pure and La-Doped $BiFeO_3$ Films Prepared by Sol–Gel Method. *Materials* **2021**, *14*, 1694. [CrossRef] [PubMed]
6. Zhang, H.; Wang, Y.; Wang, R.; Zhang, X.; Liu, C. Optimizing the Properties of $InGaZnO_x$ Thin Film Transistors by Adjusting the Adsorbed Degree of Cs^+ Ions. *Materials* **2019**, *12*, 2300. [CrossRef] [PubMed]
7. Song, W.; Jiang, B.; Ji, D. Improving the Tribological Performance of MAO Coatings by Using a Stable Sol Electrolyte Mixed with Cellulose Additive. *Materials* **2019**, *12*, 4226. [CrossRef] [PubMed]
8. Yang, K.; Zeng, J.; Huang, H.; Chen, J.; Cao, B. A Novel Self-Adaptive Control Method for Plasma Electrolytic Oxidation Processing of Aluminum Alloys. *Materials* **2019**, *12*, 2744. [CrossRef] [PubMed]
9. Yang, K.; Huang, H.; Chen, J.; Cao, B. Discharge Behavior and Dielectric Breakdown of Oxide Films during Single Pulse Anodizing of Aluminum Micro-Electrodes. *Materials* **2019**, *12*, 2286. [CrossRef] [PubMed]
10. Axelevitch, A. Hot-Probe Characterization of Transparent Conductive Thin Films. *Materials* **2021**, *14*, 1186. [CrossRef] [PubMed]

 *materials*

Article

Phase and Orientation Control of NiTiO₃ Thin Films

Jon Einar Bratvold *, Helmer Fjellvåg and Ola Nilsen

Centre for Materials Science and Nanotechnology (SMN), Department of Chemistry, University of Oslo, P.O. Box 1033 Blindern, N-0315 Oslo, Norway; helmer.fjellvag@kjemi.uio.no (H.F.); ola.nilsen@kjemi.uio.no (O.N.)
* Correspondence: j.e.bratvold@kjemi.uio.no

Received: 26 November 2019; Accepted: 20 December 2019; Published: 25 December 2019

Abstract: Subtle changes in the atomic arrangement of NiTiO₃ in the ilmenite structure affects its symmetry and properties. At high temperatures, the cations are randomly distributed throughout the structure, resulting in the corundum structure with $R-3c$ symmetry. Upon cooling, the cations order in alternating layers along the crystallographic c axis, resulting in the ilmenite structure with $R-3$ symmetry. Related to this is the $R3c$ symmetry, where the cations alternate both perpendicularly and along the c axis. NiTiO₃ with the latter structure is highly interesting as it exhibits ferroelectric properties. The close relationship between structure and properties for ilmenite-related structures emphasizes the importance of being able to control the symmetry during synthesis. We show that the orientation and symmetry of thin films of NiTiO₃ formed by atomic layer deposition (ALD) can be controlled by choice of substrate. The disordered phase ($R-3c$), previously only observed at elevated temperatures, have been deposited at 250 °C on α-Al₂O₃ substrates, while post-deposition annealing at moderate temperatures (650 °C) induces ordering ($R-3$). We have in addition explored the symmetry and epitaxial orientation obtained when deposited on substrates of LaAlO₃(100), SrTiO₃(100) and MgO(100). The presented work demonstrates the possibilities of ALD to form metastable phases through choice of substrates.

Keywords: metal oxide thin films; atomic layer deposition; ALD; crystallography; epitaxy; NiTiO₃

---

1. Introduction

NiTiO₃ (NTO) is a technologically relevant material proven suitable within photo catalysis, high-k dielectrics and non-volatile memory [1–5]. In addition, ab initio calculations predict ferroelectricity and crystal structure dependent weak ferromagnetism, potentially suitable for sensor, microwave and spintronic applications [6]. At elevated temperatures, NTO adopts the corundum structure (space group $R-3c$, Figure 1a) with random distribution of cations [7]. This disordered phase has so far been unquenchable, as cooling induces ordering of the two different cations in alternating layers perpendicular to the crystallographic c axis. As ordering occurs the c glide plane is lost, and the structure is known as the ilmenite structure (space group $R-3$, Figure 1b) [8]. This is also the case for ilmenite (FeTiO₃) itself, which additionally has been proven to adopt the related and technologically interesting LiNbO₃ structure (space group $R3c$, Figure 1c) at high temperature and pressure [9]. In the $R3c$ structure, the two different cations alternate perfectly, both parallel and perpendicular to the c axis. Distinguishing between the three different symmetries by the position of the reflections using X-ray diffraction is difficult. Unit cell values obtained from density functional theory calculations on $R-3c$ and $R3c$ are virtually the same [10], and effects from strain or off-stoichiometry are likely to be much larger. However, determining whether the symmetry is $R-3$ or $R-3c/R3c$ is easy, as the diffraction pattern from $R-3$ has reflections from ordering not allowed with the c glide plane of $R-3c/R3c$ due to extinction rules. While there are no reports of bulk NTO with $R3c$ structure, the phase was claimed to have been obtained by Varga et. al on α-Al₂O₃ substrates as thin films made by pulsed-laser deposition [10]. Even though the symmetry was not unambiguously confirmed, some degree of lattice polarization was observed, and possibly weak ferromagnetism, indicating the presence of the LiNbO$_3$ structure [11–13]. Ferroelectricity and weak ferromagnetism have indeed been observed in bulk samples of NTO as well [14], but the report did not discuss the crystal symmetry of the samples.

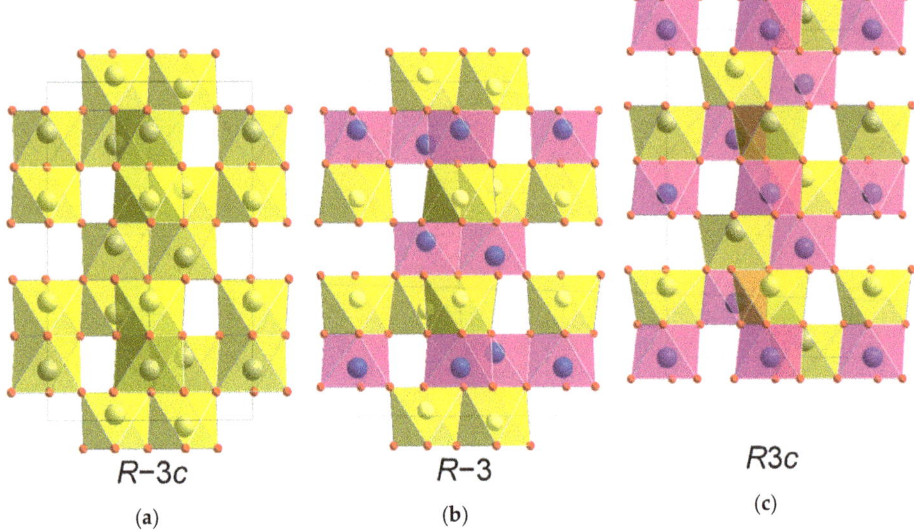

Figure 1. Schematic showing the similarity between the unit cells of space groups *R−3c* (**a**), *R−3* (**b**) and *R3c* (**c**), viewed along [110]. The *R3c* structure (**c**) is shifted up one cation layer to emphasize the relation to the two other structures.

Thin films of NTO have previously been made by various deposition techniques, including aerosol-assisted CVD [2], sol-gel methods [3,15–18], dip-coating [19] and RF-sputtering [4,5], as well as with the previously mentioned pulsed laser deposition [10–12]. Especially for the latter technique, compositional control seems to be challenging. In addition, all methods require elevated temperatures (>500 °C) either during deposition or by post-deposition annealing. While amorphous films might be desirable for some applications, a well-defined crystallinity and orientation is usually required to make use of the material. The atomic layer deposition (ALD) technique offers an alternative route for deposition of epitaxial thin films, provided careful selection of substrates [20,21], at relatively low temperatures (typically <400 °C). We have previously reported details about the deposition of NTO by ALD [22]. As deposited films on Si(100) with a 1:1 ratio between Ni and Ti (within 1%, as measured by X-ray fluorescence), all showed a preferred (001) orientation in the deposition temperature range 175–275 °C. No sign of the ordering reflections belonging to the *R−3* symmetry were visible at any temperature, neither before nor after annealing. Given the difficulty of obtaining the *R3c* phase in bulk, the films deposited on Si(100) were assumed to have the disordered *R−3c* symmetry. However, from the limited amount of data the *R3c* symmetry cannot be ruled out. In the work presented here, we show the possibility to control the orientation and crystallinity of deposited NTO films with the use of various single-crystal substrates. As with the previously reported films deposited on Si(100), the films discussed here are assumed to have the *R−3c* symmetry whenever the ordering reflections are not present. The ability to control the phase and orientation of these films is interesting with respect to catalysis. In addition, by using ALD it should be easy to incorporate other elements. Doping of NiTiO$_3$ has already been shown to improve solar water splitting efficiency [23], as well as inducing ferromagnetism in coexistence with ferroelectricity [24].

2. Materials and Methods

The films were deposited in an F-120 Sat reactor (ASM Microchemistry Ltd., Helsinki, Finland) using the precursors Ni(acac)$_2$ (nickel acetylacetonate, 95%, Sigma-Aldrich, St. Louis, MO, USA) in combination with O$_3$ (produced in a BMT Messtechnik GMBH ozone generator from 99.6% O$_2$, AGA), and TTIP (Titanium(IV) tetraisopropoxide, 97%, Sigma-Aldrich, St. Louis, MO, USA) in combination with deionized water. A pulsing and purging sequence of (2 s Ni(acac)$_2$–1.5 s purge–3 s O$_3$–2 s purge) + (0.5 s TTIP–1 s purge–2 s H$_2$O–3 s purge) were used for all depositions, as described more thoroughly in our prior work on deposition of NTO [22]. The substrates used were 3×3 cm^2 single crystals of Si(100), for thickness and composition determination, as well as α-Al$_2$O$_3$(001) (referred to simply as Al$_2$O$_3$ from now on), LaAlO$_3$(100) (LAO), SrTiO$_3$(100) (STO) and MgO(100) for structural analysis. All dust was blown clean of substrates using pressurized air before being placed in the reaction chamber, and subjected to 15 s of O$_3$ immediately before deposition. The films grown on Al$_2$O$_3$ and MgO were deposited in the same experiment with a nominal thickness of 165 nm (measured on Si(100) substrates). Likewise, the films grown on STO and LAO were deposited in the same experiment with a nominal thickness of 130 nm (measured on Si(100) substrates). All depositions were performed at 250 °C. Annealing was undertaken in air by rapid thermal processing (RTP) in an MTI Corporation OTF-1200X furnace (Richmond, VA, USA). The heating program consisted of a 15 min. ramp from room temperature to 650 °C, followed by a dwell time of 15 min., and subsequent cooling in the furnace to room temperature, for a typical duration of 30 min.

The film thicknesses were measured by spectroscopic ellipsometry on Si(100) substrates, using a J.A. Woollam (Lincoln, Dearborn, MI, USA) alpha-SE ellipsometer. The CompleteEASE software package (version 4.92, J.A. Woollam, Lincoln, Dearborn, MI, USA) was used to fit a Cauchy function to the obtained data in the 390–900 nm wavelength range. Measurements on several positions on the 3·3 cm^2 Si substrates were averaged (std.dev. ~3 nm) used to indicate a "nominal thickness" for the films on single crystalline oxide substrates. The cationic composition was determined by X-ray fluorescence (XRF) measurements on the Si substrates using a Philips (Almelo, The Netherlands) PW2400 spectrometer and the UniQuant analysis software (version 2, Omega Data Systems, Veldhoven, The Netherlands). All θ-2θ X-ray diffraction (XRD) was performed on a Bruker AXS (Karlsruhe, Germany) D8 Discover diffractometer in Bragg–Brentano configuration with Cu Kα radiation. The diffractometer was equipped with a Ge(111) monochromator and a LynxEye strip detector. Reciprocal space maps and φ scans were collected using a PANalytical Empyrean diffractometer, equipped with a Cu Kα source powered at 45 kV/40 mA, a hybrid monochromator and a PIXcel^{3D} detector (PANalytical, Almelo, The Netherlands). Atomic force microscopy (AFM) was performed with a Park Systems (Santa Clara, CA, USA) XE-70 AFM, equipped with a PPP-CONSTCR tip (Nanosensors, Neuchâtel, Switzerland) in contact mode.

Literature values for NTO in this paper refer to the PDF# 01-075-3757, ICDD. The listed cell parameters of the unit cell, having the ilmenite structure (R–3), are a = 5.0321 Å and c = 13.7924 Å.

3. Results

To provoke different crystallographic orientations, films were deposited on a range of single crystalline substrates: Al$_2$O$_3$(001), LAO(100), STO(100) and MgO(100). The results will be presented in this order below.

3.1. On Al$_2$O$_3$(001)

An obvious choice of substrate for growth of NTO thin films is Al$_2$O$_3$, as the two materials adopt closely related crystal structures. Al$_2$O$_3$ lends its mineral name to the corundum structure (R–3c), while NiTiO$_3$ has the same structure as its iron counterpart ilmenite (FeTiO$_3$), at room temperature (R–3) (Figure 1). The unit cell parameters of Al$_2$O$_3$ (a = 4.76 Å, c = 12.99 Å, PDF# 00-046-1212, ICDD) are distinctly smaller than for NTO (a = 5.03 Å, c = 13.79 Å, PDF# 01-075-3757, ICDD).

As-deposited NTO films were (00*l*) oriented on Al$_2$O$_3$(001), as observed by θ-2θ X-ray diffraction (Figure 2). Upon annealing, two additional reflections appeared: (003) and (009). These reflections are only present for the *R*−3 symmetry due to extinction rules, and can thus be used to differentiate between cation order (*R*−3) and disorder (*R*−3*c*). It should be noted that the (003) and (009) reflections are also absent for the *R3c* symmetry. However, as mentioned in the introduction, since this symmetry has never been irrefutably observed for NTO it is assumed in the following that the (00*l*) oriented films presented here have the *R*−3*c* symmetry, whenever the (003) and (009) reflections are not present.

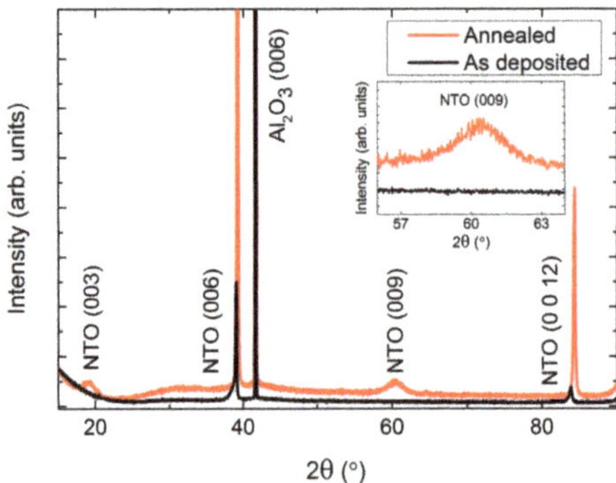

Figure 2. X-ray diffraction (XRD) of (00*l*) oriented NiTiO$_3$ (NTO) on Al$_2$O$_3$(001) showing the transition from space group *R*−3*c* as deposited (**black line**) to *R*−3, upon annealing (**red line**). The ordering reflections (003) and (009) are not visible for the as deposited films, as shown in the inset.

The crystallinity of the film improved drastically upon annealing, clearly visible from the reciprocal space map (RSM) of the NTO (006) reflection by an increase in intensity (Figure 3). A decrease in full width at half maximum (FWHM) along $q_{\parallel}$, from 1.0° to 0.7°, was also observed. Annealing shortened the *c* axis length, visible as a shift of the position of the (006) reflection to a larger q value, along $q_{\perp}$. The *c* axis was 0.4% longer in the as deposited film and 0.1% shorter in the annealed film, compared to the literature value.

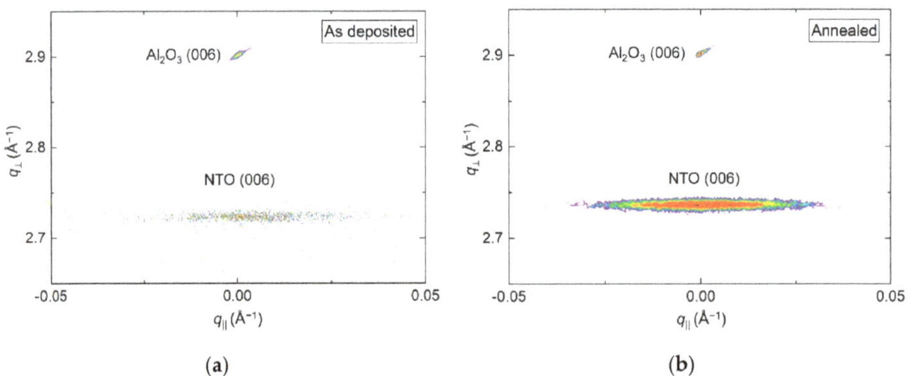

Figure 3. Reciprocal space map (RSM) of the symmetrical (006) reflections from NTO and Al$_2$O$_3$ as deposited (**a**) and after annealing (**b**).

RSM of the (1 0 10) asymmetrical reflections in Figure 4 show the in-plane relaxation of the film upon annealing. For the as deposited film, the position of the NTO (1 0 10) reflection indicated an *a* axis 0.1% shorter than the literature value. Annealing resulted in a 0.1% longer *a* axis, again, compared to the literature value. The FWHM along the Ewald sphere was also reduced upon annealing, from 1.2° to 1.0°.

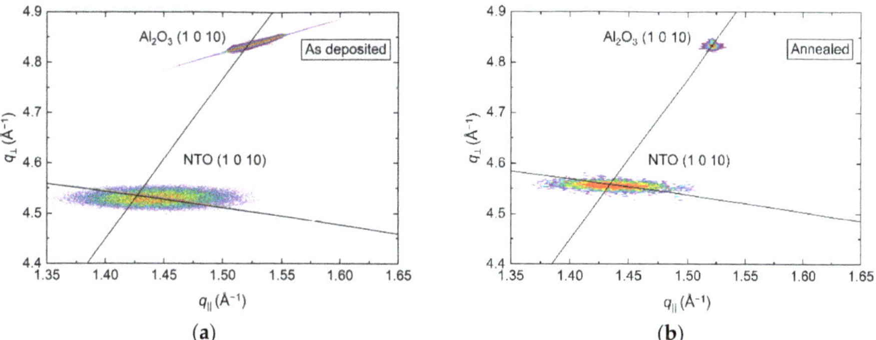

Figure 4. RSM of the asymmetrical (1 0 10) reflections from NTO and Al$_2$O$_3$ before (**a**) and after (**b**) annealing. The arced lines illustrate the Ewald sphere, while the straight lines are a guide to the eye that intersects the origin and the Al$_2$O$_3$ (1 0 10) reflection.

The φ scans of the (1 0 10) reflection revealed a six-fold rotational symmetry (Figure 5), with three of the reflections overlapping those from the substrate, and the other three shifted by 60°.

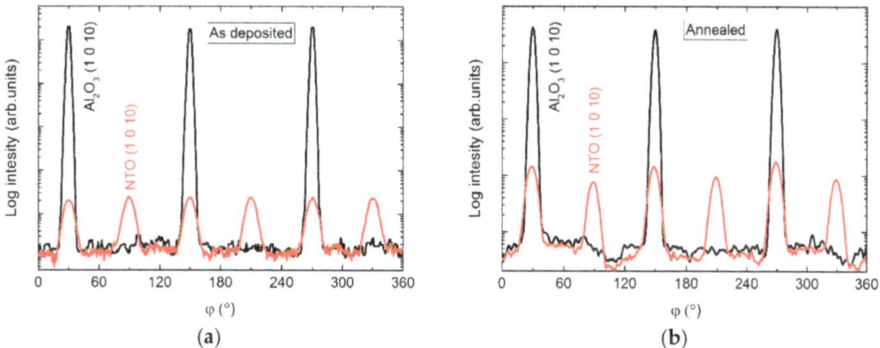

Figure 5. φ scan of the (1 0 10) reflections from NTO (**red lines**) and Al$_2$O$_3$ (**black lines**) as deposited (**a**) and after annealing (**b**).

AFM investigations of the as deposited films showed a very flat surface, with a root mean square (rms) roughness of only 0.21 nm (Figure 6). After annealing, the roughness increased only slightly, to 0.24 nm. No recognizable facets could be identified, but rather many smaller, round crystallites, possibly convoluted by the AFM tip itself.

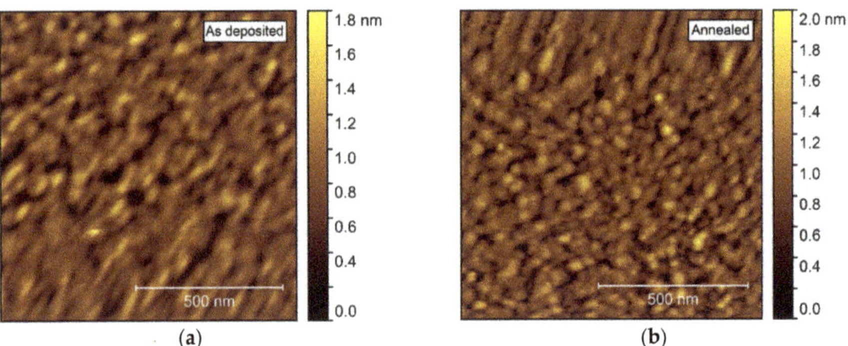

(a)　　　　　**(b)**

Figure 6. Atomic force microscopy (AFM) images of NTO deposited on Al_2O_3(001). As deposited root mean square (rms) = 0.21 nm (**a**), increasing to rms = 0.24 nm after annealing (**b**).

3.2. On LaAlO$_3$(100)

LAO has a distorted perovskite crystal structure, with space group symmetry $R-3c$, the same as the high temperature disordered phase of NTO. Compared to NTO, LAO (PDF 00-031-0022, ICDD) has a longer a axis (5.36 Å vs. 5.03 Å), but a shorter c axis (13.11 Å vs. 13.79 Å). However, LAO can also be represented as a pseudocubic structure, with angles that are less than 0.1° off from 90°. In this regard, the unit cell space group is $Pm-3m$, and $a = 3.79$ Å. From here comes the given orientation of the substrates used in this work: LAO(100), which is the same as LAO(012) in the rhombohedral system. The orientation and reflections of LAO in this text refer to the pseudocubic system, unless otherwise stated. There is no pseudocubic symmetry for NTO similar to LAO, given the same orientation as for LAO, with pseudocubic(100) = rhombohedral(012). The closest equivalent is a unit cell with dimensions $a = 5.44$ Å; $b = 5.03$ Å; $c = 3.66$ Å, and angles $\alpha = 90°$; $\beta = 83.1°$; $\gamma = 90°$.

NTO films deposited on LAO(100) substrates had two preferred orientations: (00*l*) and (*h*0*h*) (Figure 7). Reflections from the (*h*0*h*) orientation were considerably weaker and broader than from the (00*l*) orientation. In addition, annealing increased the intensity and reduced the FWHM for all reflections, with (*h*0*h*) reflections remaining distinctly broader.

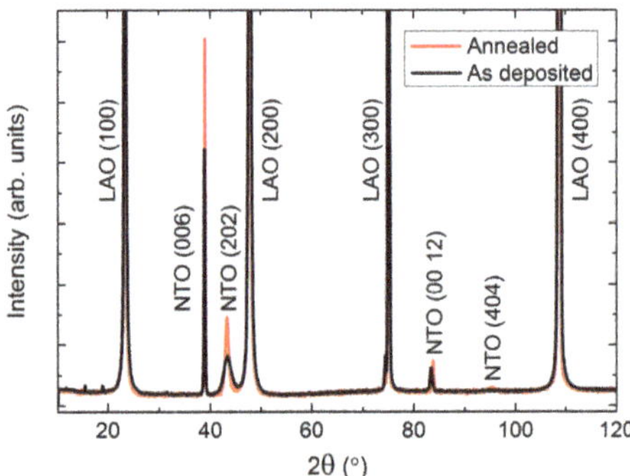

Figure 7. XRD of NTO deposited on LaAlO$_3$(100) (LAO(100)), as deposited (**black line**) and after annealing (**red line**).

RSM of the (006) reflection revealed a small shift in $q_\perp$ direction upon annealing, with the length of the c axis being 0.8% and 0.2% larger than the literature value before and after annealing, respectively (Figure 8). Annealing also reduced the FWHM from 2.3° to 2.1°, along $q_\parallel$. Extensive efforts were made to collect decent RSMs of the (*h*0*h*) reflection, but with no success. This was also the case for any related asymmetrical reflections.

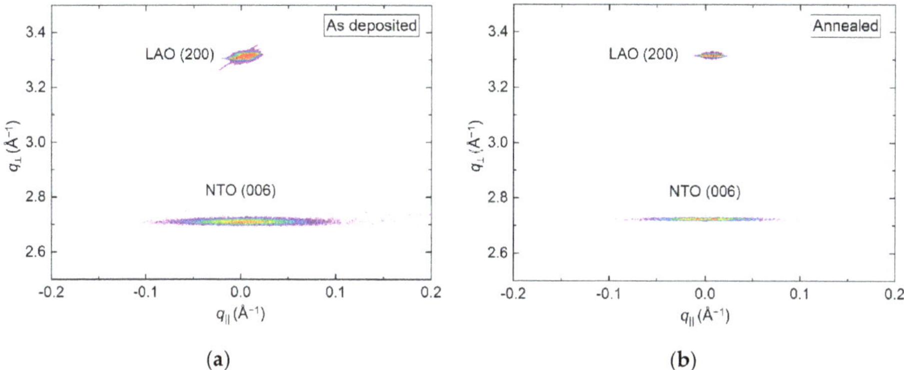

(a)

(b)

Figure 8. RSM of the symmetrical (006) reflection from NTO along with the LAO (200) reflection, before (**a**) and after (**b**) annealing.

From the (00*l*) related asymmetrical reflection (1 0 10) in Figure 9, the *a* axis was calculated to be 0.3% and 0.5% shorter compared to the literature value before and after annealing, respectively. The shift along $q_\perp$ corresponded well with the compression of the *c* axis upon annealing, and a distinct reduction of the FWHM along the Ewald sphere from 1.5° to 1.1° was also observed.

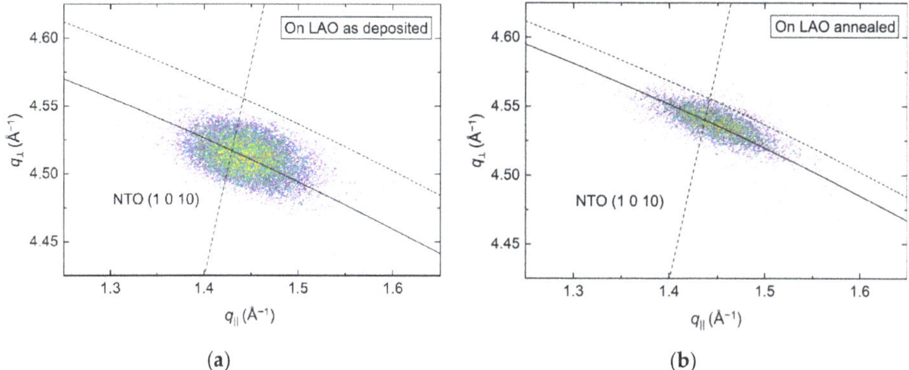

(a)

(b)

Figure 9. RSM of the asymmetrical NTO (1 0 10) reflection from a film deposited on LAO(100), as deposited (**a**) and after annealing (**b**). The solid arced lines illustrate the curvature of the Ewald sphere. The intersection of the dashed lines mark the theoretical position of the NTO (1 0 10) reflection.

The φ scans revealed a 12-fold symmetry out-of-phase with the four-fold symmetry of the selected substrate reflection (Figure 10). Annealing did not affect the symmetry, and the only observable change was a slight increase in intensity.

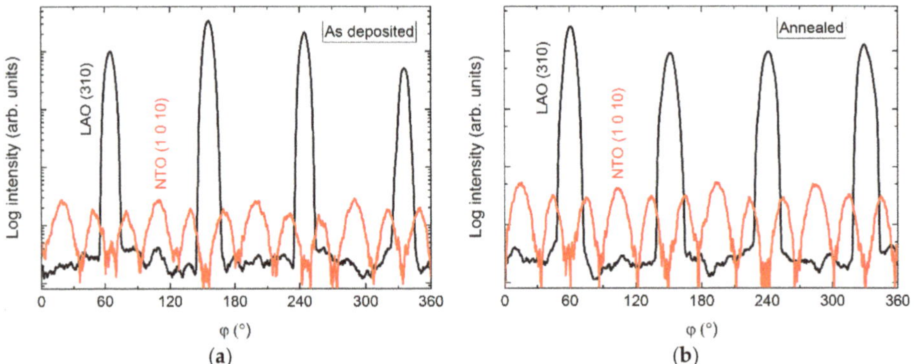

Figure 10. φ scans of the NTO (1 0 10) reflection (**red line**) along with the LAO(310) reflection (**black line**), before (**a**) and after (**b**) annealing.

Annealing induced no apparent change in the topography, as observed by AFM (Figure 11). The surface was comprised of spherical-like crystallites, similar to the films deposited on Al_2O_3, but with a larger rms roughness, at 0.9–1.0 nm.

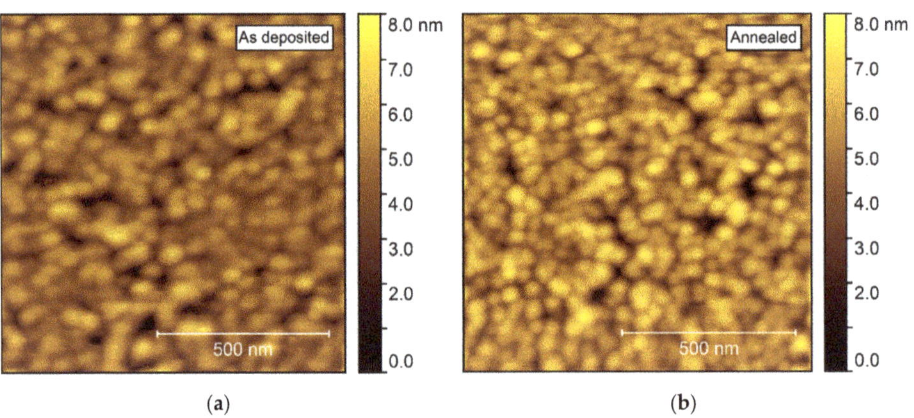

Figure 11. AFM images of NTO deposited on LAO(100). As deposited rms roughness = 0.9 nm (**a**). Annealed rms roughness = 1.0 nm (**b**).

3.3. On SrTiO₃(100)

The pseudocubic representation of LAO is closely related to STO, a perovskite structure with $Pm3m$ symmetry. The cell parameters of STO are slightly larger compared to LAO, with $a = 3.90$ Å (PDF 00-035-0734, ICDD).

Depositions on STO(100) resulted in ($h0h$) oriented NTO (Figure 12). The intensity of the reflections were higher compared to the same reflections on LAO, but the relative increase upon annealing was smaller, indicating an initial higher crystallinity of the as deposited film. The FWHM values were very similar for the two systems, both as deposited and after annealing.

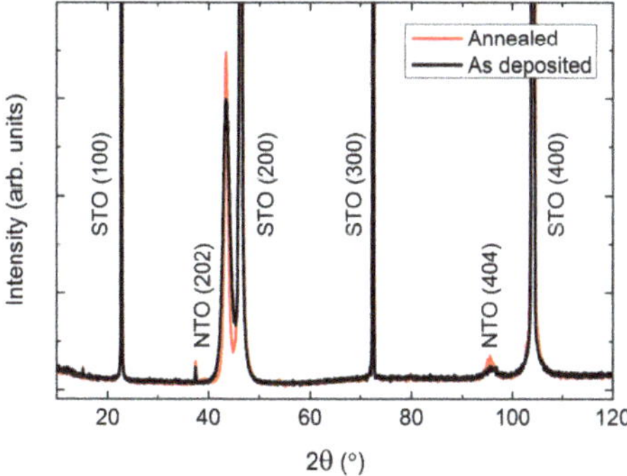

Figure 12. XRD of (*h0h*) oriented NTO deposited on SrTiO$_3$(100) (STO(100)). A slight increase in intensity was observed after annealing (**red line**), compared to the as deposited film (**black line**).

RSM of the symmetrical (202) reflection of as deposited and annealed films revealed a reduction in the FWHM from 3.4° to 2.5° along $q_\parallel$ upon annealing (Figure 13). A small shift in the $q_\perp$ direction was also observed, with the calculated c axis being 0.8% (as deposited) and 0.5% (annealed) larger than the literature value.

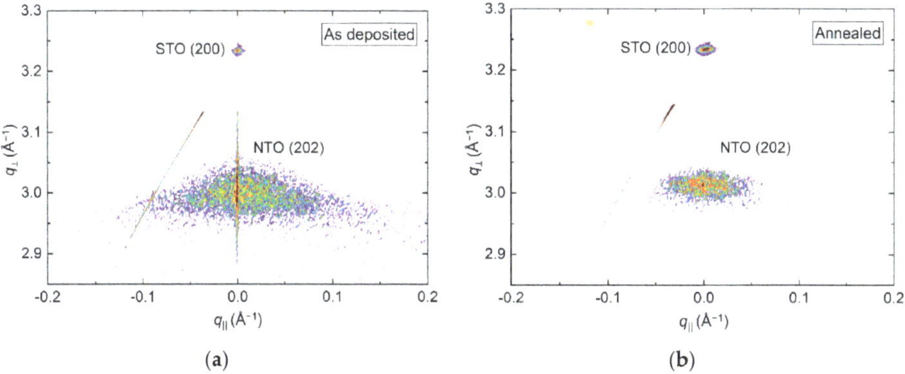

(a)   (b)

Figure 13. RSM of the symmetrical NTO(202) reflection along with STO(200) before (**a**) and after (**b**) annealing. Detector streaking is visible in both RSMs, with the as deposited data also showing wavelength streaking through the film reflection.

Conversely, investigations of the asymmetrical reflection showed a clear shift along the Ewald sphere (Figure 14), indicating a pronounced relaxation of the structure. However, a low-intensity reflection closer to the relaxed position was visible for the as-deposited film as well, indicating that the initial strain did not persist throughout the as-deposited film. A reduction in FWHM along the Ewald sphere was observed upon annealing, from 1.6° to 0.8°. As the NTO(220) reflection was very close to the edge of the accessible region of the instrument, and possibly out of reach for the as-deposited film, these values are very uncertain. This is also reflected in the calculated a axis values. Compared to literature, the a axis was 5.0% and 1.5% smaller for the as deposited and annealed films, respectively.

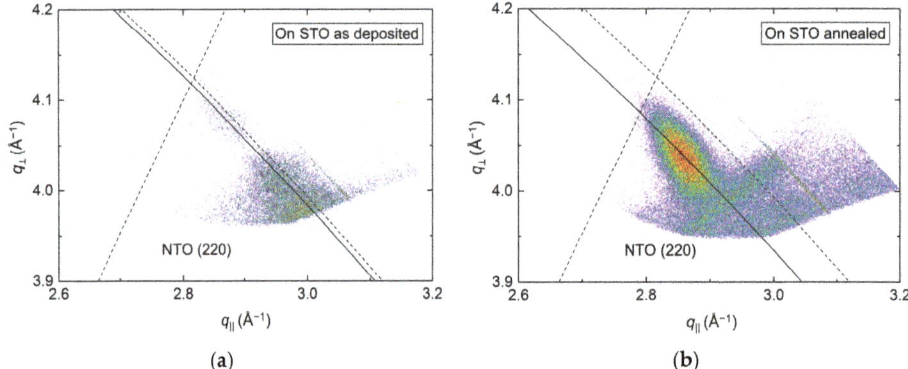

Figure 14. RSM of the asymmetrical (220) reflection from NTO on STO(100) as deposited (**a**) and after annealing (**b**). The solid arced lines illustrate the curvature of the Ewald sphere. The intersection of the dashed lines mark the theoretical position of the NTO(220) reflection.

The φ scan of the asymmetrical (220) reflection displayed a four-fold symmetry out-of-phase with the selected asymmetrical reflection from the substrate (Figure 15). Again, a slight increase in intensity was observed upon annealing.

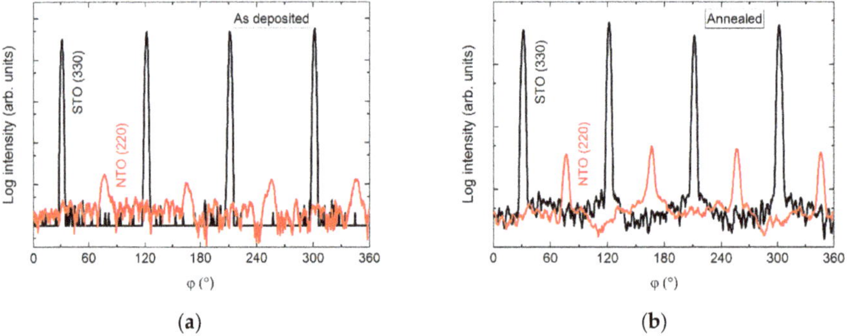

Figure 15. φ scans of the (220) reflection from NTO (red line) superimposed on the φ scan of the STO(330) reflection (black line) before (**a**) and after annealing (**b**).

Compared to films deposited on Al_2O_3, AFM of films deposited on STO revealed a considerably higher rms roughness, increasing from 1.5 to 1.7 nm upon annealing (Figure 16). As for the films deposited on Al_2O_3, the surface consisted of numerous round crystallites with no recognizable facets.

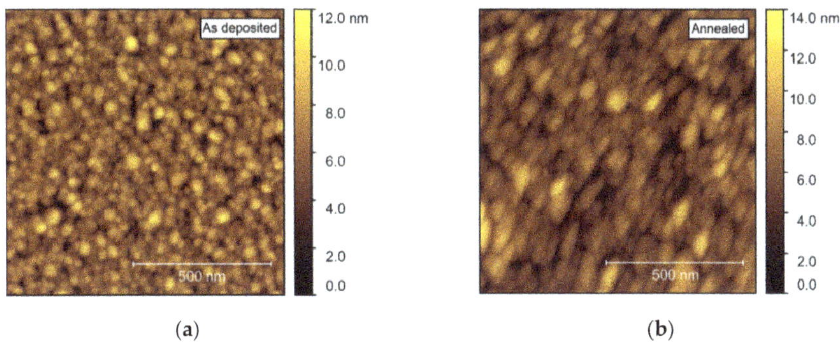

(a)   (b)

Figure 16. AFM images of NTO on STO(100), with rms roughness of 1.5 nm as deposited (**a**) and 1.7 nm after annealing (**b**).

3.4. On MgO(100)

MgO also has a cubic structure (*Fm−3m*), but with larger unit cell parameters (*a* = 4.21 Å, PDF 00-045-0946, ICDD) as compared to STO. As on STO, the deposited film was (*h0h*) oriented (Figure 17). Since the *d* spacing of NTO(*h0h*) and MgO(*h00*) are almost the same, a diffractogram of the pure MgO substrate is shown in Figure 17.

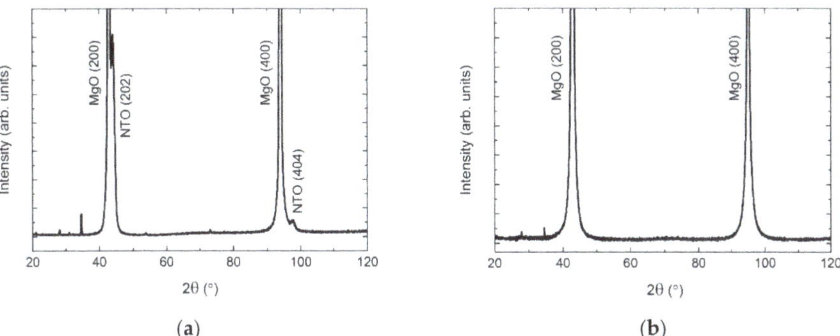

(a)   (b)

Figure 17. XRD of NTO on MgO(100) (**a**), showing the (*h0h*) orientation of the film. A diffractogram of an uncoated MgO(100) substrate is shown in (**b**) for comparison. Minor substrate artefacts are visible at $2\theta = 28°, 34°, 54°$ and $73°$.

RSM of the symmetrical (202) reflection disclosed a very sharp peak with FWHM along $q_\parallel$ of only 0.2° (Figure 18). The position of the reflection along $q_\perp$ corresponds to a unit cell with a calculated *c* axis 1.2% shorter than the literature value.

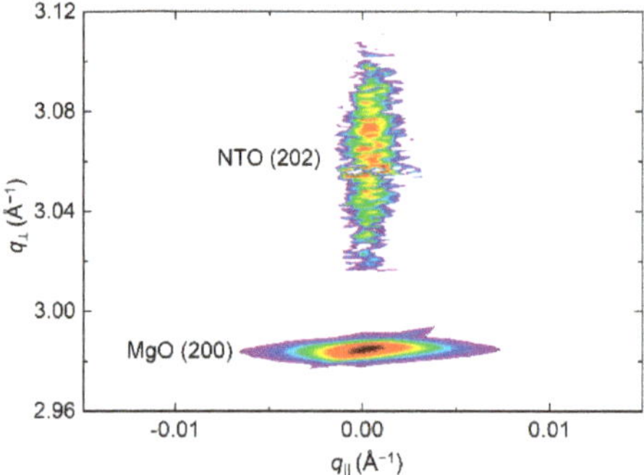

Figure 18. RSM of the symmetrical NTO (202) reflection as deposited, along with the MgO(200) reflection.

The position of the asymmetrical reflection (220) (Figure 19) was close to that observed for the annealed film deposited on STO (assumed to be relaxed). The FWHM along the Ewald sphere, at 1.5°, was slightly smaller than for as-deposited NTO on STO. The calculated a axis was 1.2% shorter compared to the literature value.

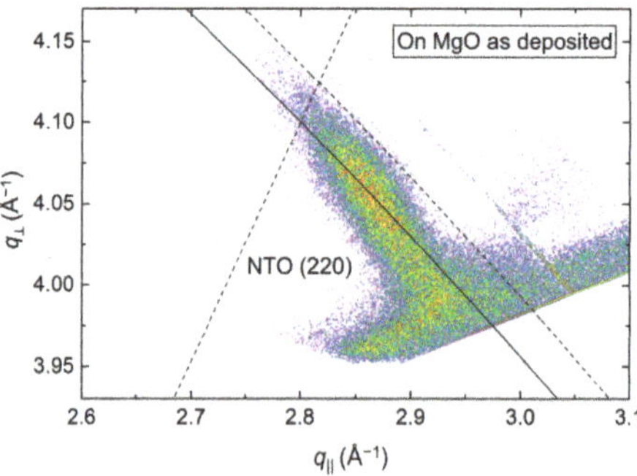

Figure 19. RSM of the asymmetrical NTO(220) reflection, as deposited on MgO(100). The Ewald sphere is illustrated as an arced solid line. The intersection of the dashed lines mark the theoretical position of the NTO(220) reflection.

In the same manner as for films on STO, the φ scan revealed a four-fold rotational symmetry of the film (Figure 20), but this time in-phase with the selected substrate reflection.

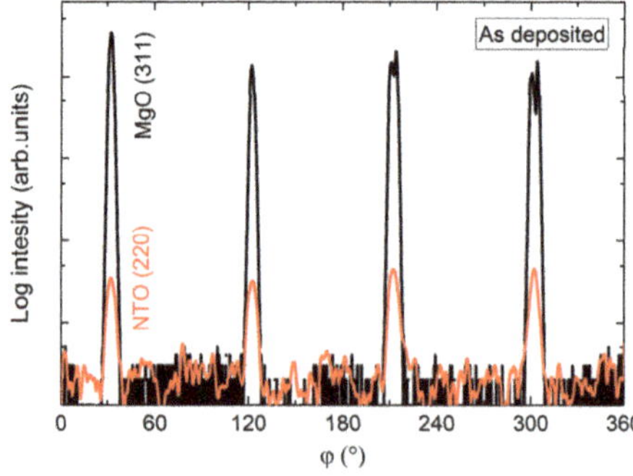

Figure 20. φ scan of the NTO(220) reflections (**red line**) superimposed on the φ scan of the MgO(311) reflections (**black line**).

Compared to the films deposited on other substrates, the surface topography of films grown on MgO was rather different. AFM scans revealed clear protrusions from a flatter bed of smaller crystals (Figure 21), with the flatter areas resembling the surfaces observed on the other substrates. The larger platelets, or flattened crystals, had no obvious collective orientation, and the rms roughness of the scanned area was 1.9 nm.

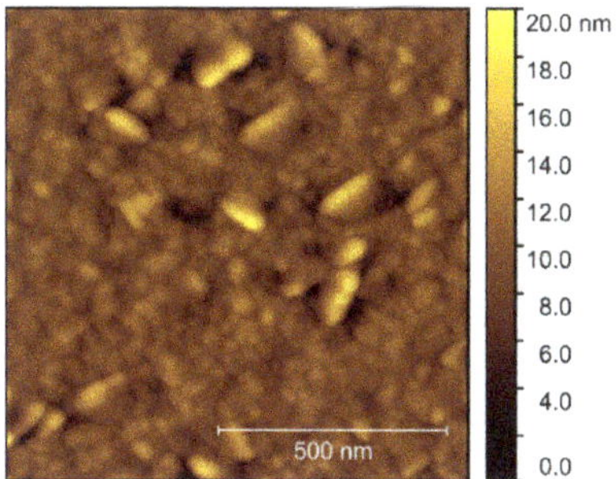

Figure 21. AFM image of NTO as deposited on MgO(100), with rms roughness = 1.9 nm.

An overview of the results is given in Table 1. All the calculated cell parameters can be found in the Supplementary Material Section S1.

Table 1. Summary of results for NTO films deposited on various single crystal substrates. AD and Ann refers to as deposited and annealed samples, respectively. The c and a values were calculated from the positions of the symmetrical and asymmetrical reflections in the RSMs, respectively, and are relative to the literature values. The ω_{FWHM} values are the full width at half maximum of Gauss functions fitted to the RSMs of symmetrical reflections along $q_\parallel$. The ES_{FWHM} values are the full width at half maximum of Gauss functions fitted to the asymmetrical reflections along the Ewald sphere. The φ values are the number of film reflections in the φ scans. Root mean square (rms) roughness values are from the built in analysis tool in the Gwyddion software.

Substrate	Film	State	c (%)	ω_{FWHM} (°)	a (%)	ES_{FWHM} (°)	φ Refl.	rms (nm)
Al$_2$O$_3$(001)	(00*l*)	AD	0.36	1.03	−0.13	1.16	6	0.21
		Ann	−0.11	0.67	0.09	0.96		0.24
LAO(100)	(00*l*)	AD	0.81	2.26	−0.29	1.49	12	0.90
		Ann	0.24	2.10	−0.45	1.06		0.96
STO(100)	(*h*0*h*)	AD	0.84	3.41	−5.01	1.61	4	1.50
		Ann	0.47	2.50	−1.52	0.77		1.71
MgO(100)	(*h*0*h*)	AD	−1.22	0.22	−1.20	1.46	4	1.93

4. Discussion

4.1. On Al$_2$O$_3$(001)

Our previous report on the growth of NTO [22] showed that films deposited on Si(100) were (00*l*) oriented, with the orientation assumed to be from an inherent preferred growth direction. Annealing of those samples improved the crystallinity, but did not alter the symmetry, as seen by the absence of the (003) and (009) reflections. On Al$_2$O$_3$, the same $R-3c$ symmetry is observed for as deposited films, but annealing induces ordering of the structure, resulting in the $R-3$ symmetry. For NTO, the disordered phase is only thermodynamically stable at temperatures above 1292 °C [7,8,25], and has not previously been obtained at room temperature by any other technique. The present case is thus yet another example of the ability of ALD to produce films with metastable phases, given careful selection of substrates and temperature treatment [26].

The mismatch factor, f, is usually given as $f = 100\% \cdot (a_s - a_f)/a_s$, where a_s and a_f denote the lattice parameters of the substrate and the film, respectively. The relevant way of looking at film-substrate mismatch in this system would be to compare the oxygen lattices at the interface. This gives a mismatch of -3.0%, when comparing the average O–O distances for the (001) surface of Al_2O_3 and NTO. The negative value of the mismatch indicates that there is a compressive strain from the substrate. Indeed, the as-deposited film has a shorter a axis, compared to the literature value. At the same time, the c axis is longer, due to the Poisson effect. Annealing relaxes the film, both in the basal plane and in the (00l) direction, with only 0.1% strain left.

The broadening of the symmetrical NTO (006) reflection along $q_{\parallel}$ and the asymmetrical NTO (1 0 10) reflection along the Ewald sphere indicates that the film consists of numerous crystallites with somewhat random texture. This could stem from abundant nucleation early in the deposition, and that the numerous crystals grow as narrow pillars throughout the film. The resulting film would have many grain boundaries and a surface with many smaller crystallites, which consequentially results in a low roughness, as indeed observed by AFM.

With an epitaxial relationship between substrate and film, one might expect the same multiplicity of reflections in the φ scan. However, as presented above, the NTO film has twice the number of reflections compared to Al_2O_3. We hypothesize that this comes from atomic steps on the substrate surface revealing trigonal surface terminations rotated at 60° from each other (Figure 22), as also seen for growth of $CaCO_3$ on Al_2O_3(001) [27]. The apparent six-fold rotational symmetry of the NTO film stems from two sets of crystallites, each with three-fold rotational symmetry, superimposed on each other. This feature also increases the likelihood of high nucleation density. The heteroepitaxial film is both out-of- and in-plane oriented, even though the number of reflections in the φ scan is different for the film and the substrate, resulting in the film‖substrate epitaxial relationship: NTO(001)[100]‖Al_2O_3(001)[100].

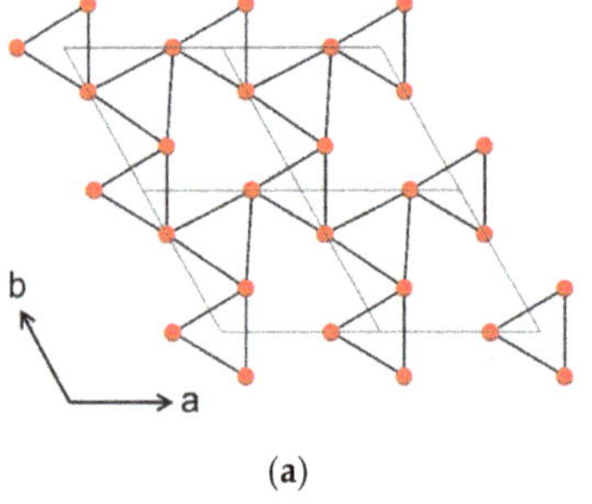

(**a**)

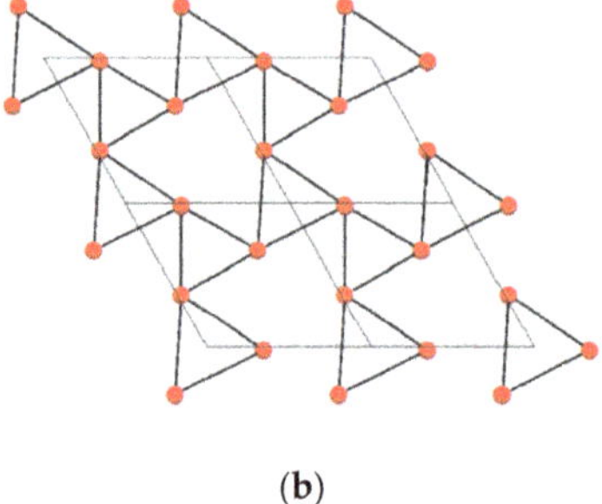

(**b**)

Figure 22. Oxygen lattices of the (001) plane in Al_2O_3, viewed along the c axis, showing the 60° rotation between the top (**a**) and bottom (**b**) faces of octahedrons around Al^{3+}. The four grey diamonds in each cartoon indicate the edges of four unit cells.

4.2. On LaAlO₃(100)

While the epitaxial relationship for NTO‖Al_2O_3 is easy to envision, the same is not true for NTO‖LAO. There are no obvious lattice matches between either of the two preferred orientations of the deposited film and the (100) surface of the LAO substrate. In addition, it is hard to untangle the nature of the ($h0h$) orientation, as it was impossible to map any related asymmetrical reflections.

With two preferred orientations the film can grow in one of three ways: either (a) the film consists of a layer with (00l) orientation close to the substrate and a ($h0h$) oriented layer on top, (b) the layers in (a) are inverted, or (c) there is a mix of crystallites with both orientations and no layered structure (Figure 23).

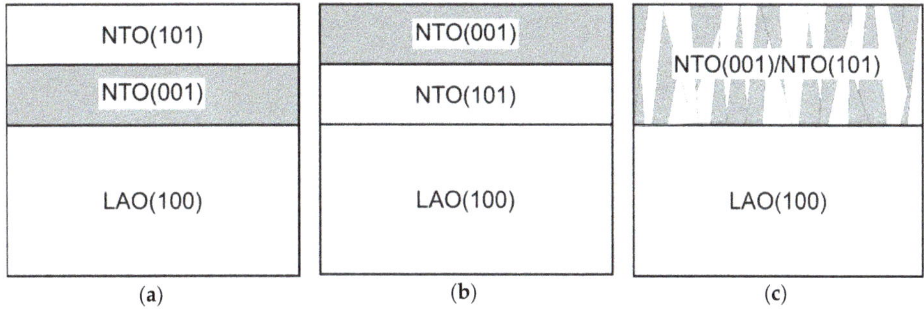

Figure 23. Cartoon of possible constellations for the two orientations observed in NTO on LAO.

We are unfortunately unable to conclude which constellation is the most probable, even with basis in the observed variations in texture and surface roughness of the films obtained in this work (see Supplementary Material Section S2 for a more thorough discussion).

The 12 NTO (1 0 10) reflections observed in the φ scan can be explained similarly to NTO on Al_2O_3. That is, the film does not actually have a 12-fold rotational symmetry, but four sets of crystallites—each set in-plane oriented, with a three-fold rotational symmetry. The orientation of the sets of crystallites are shifted by 90° with respect to each other, resulting in 12 reflections in the φ scan. Both the substrate and a NTO(*h0h*) layer could facilitate this. In any case, it is clear that the nucleation density is high, as the different sets of NTO(00*l*) crystallites must have nucleated separately. If the NTO(00*l*) crystallites nucleated on the LAO(100) surface, as in constellation (a) or (c), the epitaxial relationship with the substrate would be: NTO(001)[100]‖LAO(100)⟨010⟩, with ⟨010⟩ representing the four identical [010] directions on the LAO(100) surface.

4.3. On SrTiO₃(100)

As with NTO on LAO, there are no obvious lattice matches between the film and the substrate. Still, the deposited film shows an in-plane (*h0h*) orientation. The four asymmetrical reflections from the film are perfectly out of phase with the four reflections from the substrate. This means that the direction of the NTO(220) scattering vector, projected down to the substrate surface, is 45° off from the same projection of the STO(330) scattering vector. Figure 24 shows the STO(100) and NTO(101) oxygen lattices, with the film lattice rotated according to the projections of the asymmetrical scattering vectors (see Supplementary Material Section S3 for details). The average O–O distance of the film lattice perpendicular to the projection of the NTO(220) direction is close to the oxygen spacing of STO(100). The mismatch along one row of NTO oxygen atoms is only −0.6%. However, when considering the direction along the NTO(220) scattering vector the mismatch between every second layer of oxygen atoms in STO(100) compared to every layer in NTO(101) is 18.4%.

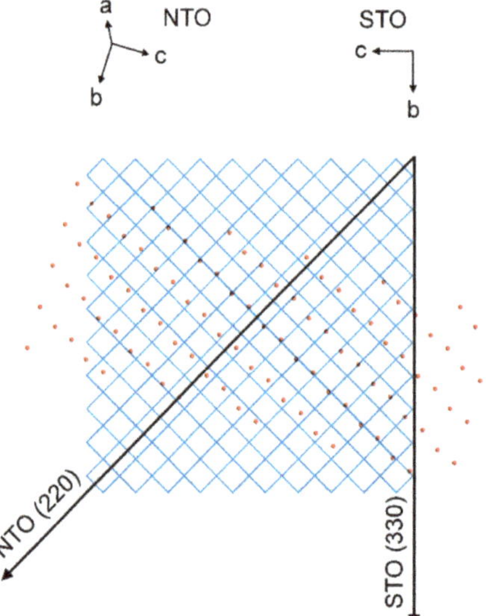

Figure 24. Schematic showing the direction of the STO(330) scattering vector projected down on the oxygen lattice of the STO(100) plane, represented by the corners of the blue matrix. Superimposed is the direction of the NTO(220) scattering vector projected down on the substrate plane, and the oxygen lattice of the NTO(101) plane, represented by red circles.

The number of reflections in the φ scan can be explained in the same way as for NTO on Al_2O_3 and LAO. A single crystalline, ($h0h$) oriented, NTO film would have only one ($hh0$) reflection in a φ scan. The STO(100) surface has four identical (010) directions: (010), ($0\bar{1}0$). (001) and ($00\bar{1}$), resulting in the four-fold rotational symmetry seen in the φ scan. As the NTO film does not nucleate as a continuous layer, the initial islands of ($h0h$) oriented crystallites do not differentiate between the STO⟨010⟩ directions. Some of the crystallites will be in-plane oriented with the STO(010) direction, and some with the other directions. An epitaxial relationship between film and substrate can thus be described as: NTO(101)[220]∥STO(100)⟨010⟩.

4.4. On MgO(100)

Unlike the previous cases, AFM of NTO on MgO show clear signs of bigger crystals. This resonates well with the smaller FWHM of the symmetrical NTO(202) reflection, indicating the presence of larger and/or less tilted crystallites. However, again, there are no obvious lattice matches between the MgO(100) surface and NTO, neither in general, nor when comparing the NTO(101) oxygen lattice with the oxygen lattice of the MgO(100) surface. Nevertheless, the film is indeed in-plane orientated with the substrate, but in the same manner as for NTO on STO. The φ scan reflections from the substrate and film are overlapping. This means that the direction of the asymmetrical scattering vectors, when projected down on the MgO(100) surface plane, are pointing in the same direction (Figure 25). Regarding a row of NTO oxygen atoms perpendicular to the projection of the MgO(311) scattering direction, there is a seemingly good lattice match for every second layer of oxygen atoms. The calculated mismatch is only 2.4%, but as can be seen from Figure 25, this is only true for selected rows of NTO oxygen atoms. When considering the whole oxygen lattice of NTO it does not correspond very well with the MgO(100) oxygen lattice. Still, the film is in-plane oriented with all the identical MgO⟨011⟩ directions, and the epitaxial relationship is determined to be: NTO(101)[220]∥MgO(100)⟨011⟩.

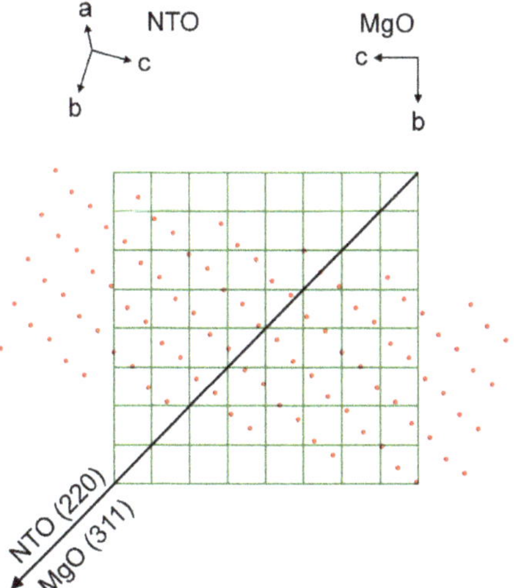

Figure 25. Schematic showing the direction of the MgO(311) scattering vector projected down on the oxygen lattice of the MgO(100) plane, represented by the corners of the green matrix, which also corresponds to the unit cell of MgO. Superimposed is the direction of the NTO(220) scattering vector projected down on the substrate plane, and the oxygen lattice of the NTO(202) plane, represented by red circles.

5. Conclusions

Oriented NTO films were successfully deposited at 250 °C by ALD on various single crystalline substrates. All films had preferred orientations as deposited. Crystallinity improved after annealing at 650 °C for 15 min, but did not change the preferred orientations. On $Al_2O_3(001)$, annealing led to cation ordering, and a change of symmetry from $R-3c$ to $R-3$, visible by the appearance of the (003) and (009) reflections. However, the intensity was much lower, and the FWHM much larger, than for the other (00*l*) reflections, indicating that only parts of the film underwent cation ordering. AFM revealed very low rms roughnesses of about 0.2 nm, both for the as deposited and annealed film. The film‖substrate epitaxial relationship was determined to be NTO(001)[100]‖Al_2O_3(001)[100]. Depositions on LAO(100) resulted in films with two preferred orientations: NTO(*h0h*) and NTO(00*l*). There were no signs of ordering reflections for either orientation, indicating a $R-3c$ symmetry. Annealing improved the crystallinity slightly, but did not alter the orientation of the film, the cation ordering nor change the rms roughness significantly, remaining at approximately 1 nm. Films deposited on STO(100) had a NTO(101)[220]‖STO(100)⟨010⟩ epitaxial relationship with the substrate, with no signs or ordering reflections Annealing resulted in a slight improvement of crystallinity and increase in roughness, from 1.5 to 1.7 nm. Films deposited on MgO(100) had $R-3c$ symmetry and the film‖substrate epitaxial relationship was NTO(101)[220]‖MgO(100)⟨011⟩. Reciprocal space mapping of the symmetrical NTO (202) reflection revealed a very small FWHM along $q_∥$, indicating large crystallites in the film, and/or a very low tilt of the crystallites. No annealing was performed, but AFM investigations displayed larger crystallites protruding from a flatter surface, and the rms roughness was significantly larger than for films on other substrates, at 1.9 nm.

Supplementary Materials: Supplementary material is available online at http://www.mdpi.com/1996-1944/13/1/112/s1.

Author Contributions: Conceptualization, J.E.B., H.F. and O.N.; Data curation, J.E.B.; Formal analysis, J.E.B. and O.N.; Funding acquisition, H.F. and O.N.; Investigation, J.E.B.; Methodology, J.E.B. and O.N.; Project administration, H.F. and O.N.; Supervision, H.F. and O.N.; Validation, O.N.; Visualization, J.E.B.; Writing—original draft, J.E.B. and O.N.; Writing—review and editing, J.E.B. and O.N. All authors have read and agreed to the published version of the manuscript.

Funding: This research was funded by University of Oslo.

Acknowledgments: The authors would like to acknowledge Henrik Hovde Sønsteby and Øystein Slagtern Fjellvåg (University of Oslo) for help with obtaining RSMs and discussions in the related analysis, and the department of Geology (University of Oslo) for use of the XRF equipment.

Conflicts of Interest: The authors declare no conflict of interest.

References

1. Yuan, P.-H.; Fan, C.-M.; Ding, G.-Y.; Wang, Y.-F.; Zhang, X.-C. Preparation and photocatalytic properties of ilmenite $NiTiO_3$ powders for degradation of humic acid in water. *Int. J. Miner. Metall. Mater.* **2012**, *19*, 372–376. [CrossRef]
2. Tahir, A.A.; Mazhar, M.; Hamid, M.; Wijayantha, K.G.; Molloy, K.C. Photooxidation of water by $NiTiO_3$ deposited from single source precursor $[Ni_2Ti_2(OEt)_2(\mu\text{-}OEt)_6(acac)_4]$ by AACVD. *Dalton Trans.* **2009**, *19*, 3674–3680. [CrossRef] [PubMed]
3. Chuang, S.-H.; Hsieh, M.-L.; Wu, S.-C.; Lin, H.-C.; Chao, T.-S.; Hou, T.-H. Fabrication and Characterization of High-*k* Dielectric Nickel Titanate Thin Films Using a Modified Sol-Gel Method. *J. Am. Ceram. Soc.* **2011**, *94*, 250–254. [CrossRef]
4. Inamdar, A.I.; Kim, J.; Jang, B.; Kim, D.; Im, H.; Jung, W.; Kim, H. Memory Conductance Switching in a Ni–Ti–O Compound Thin Film. *Jpn. J. Appl. Phys.* **2012**, *51*, 104102. [CrossRef]
5. Ruiz-Preciado, M.A.; Bulou, A.; Makowska-Janusik, M.; Gibaud, A.; Morales-Acevedo, A.; Kassiba, A. Nickel titanate ($NiTiO_3$) thin films: RF-sputtering synthesis and investigation of related features for photocatalysis. *CrystEngComm* **2016**, *18*, 3229–3236. [CrossRef]
6. Xin, C.; Wang, Y.; Sui, Y.; Wang, Y.; Wang, X.; Zhao, K.; Liu, Z.; Li, B.; Liu, X. Electronic, magnetic and multiferroic properties of magnetoelectric $NiTiO_3$. *J Alloys Compd.* **2014**, *613*, 401–406. [CrossRef]
7. Jacob, K.T.; Saji, V.S.; Reddy, S.N.S. Thermodynamic evidence for order–disorder transition in $NiTiO_3$. *J. Chem. Thermodyn.* **2007**, *39*, 230–235. [CrossRef]
8. Lerch, M.; Boysen, H.; Neder, R.; Frey, F.; Laqua, W. Neutron scattering investigation of the high temperature phase transition in nickel titanium oxide ($NiTiO_3$). *J. Phys. Chem. Solids* **1992**, *53*, 1153–1156. [CrossRef]
9. Ming, L.C.; Kim, Y.-H.; Uchida, T.; Wang, Y.; Rivers, M. In situ X-ray diffraction study of phase transitions of $FeTiO_3$ at high pressures and temperatures using a large-volume press and synchrotron radiation. *Am. Mineral.* **2006**, *91*, 120–126. [CrossRef]
10. Varga, T.; Droubay, T.C.; Bowden, M.E.; Nachimuthu, P.; Shutthanandan, V.; Bolin, T.B.; Shelton, W.A.; Chambers, S.A. Epitaxial growth of $NiTiO_3$ with a distorted ilmenite structure. *Thin Solid Films* **2012**, *520*, 5534–5541. [CrossRef]
11. Varga, T.; Droubay, T.C.; Bowden, M.E.; Colby, R.J.; Manandhar, S.; Shutthanandan, V.; Hu, D.; Kabius, B.C.; Apra, E.; Shelton, W.A.; et al. Coexistence of weak ferromagnetism and polar lattice distortion in epitaxial $NiTiO_3$ thin films of the $LiNbO_3$-type structure. *J. Vac. Sci. Technol. B* **2013**, *31*. [CrossRef]
12. Varga, T.; Droubay, T.C.; Bowden, M.E.; Stephens, S.A.; Manandhar, S.; Shutthanandan, V.; Colby, R.J.; Hu, D.; Shelton, W.A.; Chambers, S.A. Strain-dependence of the structure and ferroic properties of epitaxial $Ni_{1-x}Ti_{1-y}O_3$ thin films grown on sapphire substrates. *Thin Solid Films* **2015**, *578*, 113–123. [CrossRef]
13. Varga, T.; Droubay, T.C.; Kovarik, L.; Hu, D.; Chambers, S.A. Controlling the structure and ferroic properties of strained epitaxial $NiTiO_3$ thin films on sapphire by post-deposition annealing. *Thin Solid Films* **2018**, *662*, 47–53. [CrossRef]
14. Acharya, T.; Choudhary, R.N.P. Structural, Ferroelectric, and Electrical Properties of $NiTiO_3$ Ceramic. *J. Electron. Mater.* **2014**, *44*, 271–280. [CrossRef]
15. Phani, A.R.; Santucci, S. Microwave irradiation as an alternative source for conventional annealing: A study of pure TiO_2, $NiTiO_3$, $CdTiO_3$ thin films by a sol–gel process for electronic applications. *J. Phys. Condens. Matter* **2006**, *18*, 6965–6978. [CrossRef]
16. Phani, A.R.; Santucci, S. Structural characterization of nickel titanium oxide synthesized by sol–gel spin coating technique. *Thin Solid Films* **2001**, *396*, 1–4. [CrossRef]

17. Taylor, D.J.; Fleig, P.F.; Schwab, S.T.; Page, R.A. Sol-gel derived, nanostructured oxide lubricant coatings. *Surf. Coat. Technol.* **1999**, *120–121*, 465–469. [CrossRef]

18. Mohammadi, M.R.; Fray, D.J. Mesoporous and nanocrystalline sol–gel derived $NiTiO_3$ at the low temperature: Controlling the structure, size and surface area by Ni:Ti molar ratio. *Solid Stat. Sci.* **2010**, *12*, 1629–1640. [CrossRef]

19. Ortiz de Zárate, D.; Boissière, C.; Grosso, D.; Albouy, P.-A.; Amenitsch, H.; Amoros, P.; Sanchez, C. Preparation of multi-nanocrystalline transition metal oxide (TiO_2–$NiTiO_3$) mesoporous thin films. *New J. Chem.* **2005**, *29*, 141–144. [CrossRef]

20. Miikkulainen, V.; Leskelä, M.; Ritala, M.; Puurunen, R.L. Crystallinity of inorganic films grown by atomic layer deposition: Overview and general trends. *J. Appl. Phys.* **2013**, *113*, 2. [CrossRef]

21. Faugier-Tovar, J.; Lazar, F.; Marichy, C.; Brylinski, C. Influence of the Lattice Mismatch on the Atomic Ordering of ZnO Grown by Atomic Layer Deposition onto Single Crystal Surfaces with Variable Mismatch (InP, GaAs, GaN, SiC). *Condens. Matter* **2017**, *2*, 3. [CrossRef]

22. Bratvold, J.E.; Fjellvåg, H.; Nilsen, O. Atomic Layer Deposition of oriented nickel titanate ($NiTiO_3$). *Appl. Surf. Sci.* **2014**, *311*, 478–483. [CrossRef]

23. Munawar, K.; Perveen, F.; Shahid, M.M.; Basirun, W.J.; Bin Misran, M.; Mazhar, M. Synthesis, characterization and computational study of an ilmenite-structured $Ni_3Mn_3Ti_6O_{18}$ thin film photoanode for solar water splitting. *New J. Chem.* **2019**, *43*, 11113–11124. [CrossRef]

24. Tursun, R.; Su, Y.C.; Yu, Q.S.; Tan, J.; Hu, T.; Luo, Z.B.; Zhang, J. Effect of doping on the structural, magnetic, and ferroelectric properties of $Ni_{1-x}A_xTiO_3$ (A = Mn, Fe, Co, Cu, Zn; x = 0, 0.05, and 0.1). *J. Alloys Compd.* **2019**, *773*, 288–298. [CrossRef]

25. Boysen, H.; Frey, F.; Lerch, M.; Vogt, T. A neutron powder investigation of the high-temperature phase transition in $NiTiO_3$. *Z. Kristallogr.* **1995**, *210*, 328–337. [CrossRef]

26. Sønsteby, H.H.; Bratvold, J.E.; Weibye, K.; Fjellvåg, H.; Nilsen, O. Phase Control in Thin Films of Layered Cuprates. *Chem. Mater.* **2018**, *30*, 1095–1101. [CrossRef]

27. Nilsen, O.; Fjellvåg, H.; Kjekshus, A. Growth of calcium carbonate by the atomic layer chemical vapour deposition technique. *Thin Solid Films* **2004**, *450*, 240–247. [CrossRef]

Article

Atomic Layer Deposition of GdCoO$_3$ and Gd$_{0.9}$Ca$_{0.1}$CoO$_3$

Marion Duparc, Henrik Hovde Sønsteby, Ola Nilsen, Anja Olafsen Sjåstad and Helmer Fjellvåg *

Centre for Materials Science and Nanotechnology, Department of Chemistry, University of Oslo, 0315 Oslo, Norway; m.j.l.duparc@smn.uio.no (M.D.); h.h.sonsteby@kjemi.uio.no (H.H.S.); ola.nilsen@kjemi.uio.no (O.N.); a.o.sjastad@kjemi.uio.no (A.O.S.)
* Correspondence: helmer.fjellvag@kjemi.uio.no

Received: 16 November 2019; Accepted: 16 December 2019; Published: 19 December 2019

Abstract: Thin films of the catalytically interesting ternary and quaternary perovskites GdCoO$_3$ and Gd$_{0.9}$Ca$_{0.1}$CoO$_3$ are fabricated by atomic layer deposition using metal β-diketonates and ozone as precursors. The resulting thin films are amorphous as deposited and become single-oriented crystalline on LaAlO$_3$(100) and YAlO$_3$(100/010) after post-annealing at 650 °C in air. The crystal orientations of the films are tunable by choice and the orientation of the substrate, mitigated through the interface via solid face epitaxy upon annealing. The films exhibit no sign of Co^{2+}. Additionally, high-aspect-ratio Si(100) substrates were used to document the suitability of the developed process for the preparation of coatings on more complex, high-surface-area structures. We believe that coatings of GdCoO$_3$ and Gd$_{1-x}$Ca$_x$CoO$_3$ may find applications within oxidation catalysis.

Keywords: gadolinium cobaltites; atomic layer deposition; β-diketonates; ozone; preferential crystal growth orientation; high-aspect-ratio substrate

---

1. Introduction

Rare-earth element perovskites with the formula ABO$_3$ (A = alkaline/ rare-earth element, B = 3d–5d transition metal) have received much attention in the field of heterogeneous catalysis [1–3]. The catalytic activity of these materials relates to the nature of the B-site element [2]. In addition, the partial substitution of the A-site alkaline/rare-earth element with a lower valency cation (typically Ca or Sr) may result in oxygen nonstoichiometry, which in turn induces specific effects on the catalytic performance [3]. Encouraging results for catalytic oxidation reactions have been obtained with LaCoO$_3$ and La$_{1-x}$A'$_x$CoO$_3$ (A' = Ca or Sr) [1,3]. However, the basicity of lanthanum makes such catalysts vulnerable to detrimental volume expansion due to lanthanum oxide hydration upon reaction with air and moisture [4,5]. Preliminary bench-scale catalyst performance tests of bulk Gd$_{1-x}$Ca$_x$CoO$_3$ for ammonia oxidation show comparable catalytic performance to the corresponding La-based system, but without the undesired degradation of the catalysts due to hydroxide formation upon temperature cycling in the processing atmosphere [6]. We currently focus on GdCoO$_3$-based catalysts of relevance for the ammonia slip reaction (i.e., the oxidation of minute quantities of NH$_3$ in a process stream into nitrogen and steam), owing to the lower basicity, and thus improved resistance towards hydration, of such Gd-containing compounds in realistic processing environments [7,8]. Notably, we also explore deposition routes for Ca-substituted variants, providing means for oxygen vacancies.

Recent literature underlines the pertinence of using atomic layer deposition (ALD) in the design and study of coatings for heterogeneous catalysis [9–11]. The sequential nature of the ALD technique inherently rules out any gas phase reactions, and the self-limiting nature of the processes leads to controllable and reproducible synthesis of morphologically and chemically uniform materials [12–14]. A major advantage of ALD compared to conventional thin-film synthesis routes like sputtering or CVD is the possibility of obtaining high-surface-area supported catalysts by depositing chemically uniform thin films of the active phase on a high-surface-area support [15,16]. This is enabled by the self-limiting mechanism that allows for deposition beyond the line-of-sight.

ALD processes have been developed and reported for a wide variety of oxides, including around 30 functional perovskites [17,18]. The development of ALD processes for ternary and quaternary oxides has recently gained attention due to their high potential in a range of applications, such as ferroelectrics, photovoltaics, and battery technology [18,19]. However, owing to the complexity of multi-cation deposition, the available ALD processes for quaternary oxides are still limited to a few systems [20]. To the best of our knowledge, no reports have been published on the preparation of ALD films of $GdCoO_3$ or the substituted variants thereof.

ALD of $LaCoO_3$ using β-diketonates and ozone was reported in 1997 by Seim et al. [21]. No reports were made of any structural or functional characterization of the product films. More recently, ALD of the quaternary $La_{1-x}Sr_xCoO_{3-\delta}$ system for the composition range $0.3 < x < 0.7$ was achieved by Ahvenniemi et al., using the same type of process [20]. One of the challenges of introducing cobalt in complex oxide ALD is catalytic decomposition of ozone and the metal-organic precursors by CoO_x species. This challenge can be overcome by tuning the precursor flux and precursor sequence, similar to recent reports on the deposition of lanthanum cuprate, for which CuO_x species exhibit the same detrimental catalytic precursor decomposition [22].

In this work we report for the first time the controlled thin-film growth and characterization of the ternary $GdCoO_3$ and quaternary $Gd_{1-x}Ca_xCoO_3$ rare-earth cobaltites using β-diketonates and ozone as precursors on flat and high-aspect ratio substrates. The current investigation is a step towards the growth and tailoring of highly selective complex thin films for heterogeneous catalysis.

2. Experimental

2.1. ALD and Precursors

All thin films were deposited in a F-120 Sat reactor (ASM Microchemistry, Helsinki, Finland) at a reactor temperature of 300 °C, unless otherwise stated. The temperature was chosen to comply with applicable temperatures for the binary oxide processes. Nitrogen was used as a purging gas, supplied from gas cylinders (99.999%, Praxair Norway, Oslo, Norway) and run through a Mykrolis purifier (Avantor Fluid Handling LLC, Devens, MA, USA) to remove oxygen and water impurities. The purging gas was maintained at a $300 \, cm^3 \, min^{-1}$ flow rate, giving an operating pressure of 2.6 mbar throughout the process.

$Co(thd)_2$ (99.9+%, Volatec, Porvoo, Finland), $Gd(thd)_3$ (99.9%, Strem Chemicals Inc., Kehl, Germany) and $Ca(thd)_2$ (99.9+%, Volatec) were used as cation sources, maintained in open boats in the reactor at 115 °C, 140 °C, and 198 °C, respectively (thd = 2,2,6,6-tetramethyl-3,5-heptanedionate). All precursors were re-sublimated before use to enhance purity. O_3 was used as the oxygen source, made from O_2 gas (99.6%, Praxair Norway) with an In USA (AC-2505) ozone generator producing 15 mass% O_3 in O_2. Pulse durations were set to 1.5 s for all metal precursors, whereas ozone was pulsed for 5 s subsequent to $Co(thd)_2$ and 3 s subsequent to $Gd(thd)_3$ and $Ca(thd)_2$ pulses. All purge durations were set to 2 s. The pulse and purge durations were chosen in agreement with previous reports of self-limiting growth using these precursors in similar reactor infrastructures.

2.2. Substrates and Annealing

Films were routinely deposited on $1 \times 1 \, cm^2$ Si(100) for characterization of thickness, whereas $3 \times 3 \, cm^2$ Si(100) substrates were used for compositional analysis with X-ray fluorescence (XRF) and for investigating any thickness gradients. Selected films were deposited on $1 \times 1 \, cm^2$ $LaAlO_3$ (100)$_{pseudocubic}$ (LAO, MTI Corp., Richmond, CA, USA), $YAlO_3$ (100) (MTI Corp.) and $YAlO_3$ (010) (YAP, MTI Corp.) for the facilitation of epitaxial growth. The investigation of conformality on high-aspect-ratio substrates was carried out on silicon substrates with parallel grooves of 20 μm depth and 10 μm width (SINTEF IKT made by reactive ion etching with the Bosch process).

The selected films were annealed at 650 °C for 30 min in 1 atm air in an OTF-1200X rapid thermal processing (RTP) furnace (MTI Corp., Richmond, CA, USA) to facilitate crystallization prior to structural investigation.

2.3. Characterization

Film thickness was routinely studied using a J. A. Woollam alpha-SE spectroscopic ellipsometer (J.A. Woollam Co., Lincoln, NE, USA) in the wavelength range 390–900 nm. A Cauchy function was successfully used to model the collected data.

X-ray diffraction (XRD) measurements were used to investigate the out-of-plane crystalline orientation of the thin films on single crystal substrates. Symmetric θ-2θ-scans were carried out on a Bruker AXS D8 Discover diffractometer (Bruker AXS, Karlsruhe, Germany) equipped with a LynxEye strip detector (Bruker AXS) and a Ge (111) focusing monochromator, providing $CuK\alpha_1$ radiation.

Chemical composition was analyzed using a Panalytical Axios Max Minerals XRF system (Malvern Panalytical, Malvern, UK) equipped with a 4 kW Rh tube. Omnian and Stratos options were employed for standardless measurements of thin film cation composition.

The chemical state of the cations, particularly cobalt, was investigated by X-ray photoelectron spectrometry (XPS) using a Thermo Scientific Theta Probe Angle-Resolved XPS system (ThermoFisher Scientific, Waltham, MA, USA). The instrument was run with a standard Al Kα source (hν = 1486.6 eV), and the analysis chamber pressure was maintained on the order of 10^{-8} mbar. Pass energy values of 200 eV and 50 eV were employed for survey scans and detailed scans, respectively. The data were corrected for any drift by setting the binding energy for adventitious carbon to 284.8 eV. Data treatment and fitting were performed within the Avantage software suite (ThermoFisher Scientific). The background was fitted to a Shirley-type pseudostep function.

Cross section SEM images of the deposited films were obtained using a Hitachi SU8230 SEM (Hitachi, Krefeld, Germany) with a cold cathode field emission electron gun. The total voltage was set to 2 kV and the films were imaged by means of secondary and back-scattered electrons.

3. Results and Discussion

3.1. Deposition of GdCoO$_3$

The development of ternary deposition processes typically requires insight into the individual growth behavior of the binary components. ALD of Co_3O_4, using $Co(thd)_2$ as a precursor and ozone as the oxidizing agent, was established by Klepper et al. in the 114–307 °C temperature range, with a growth per cycle (GPC) of $\approx$ 0.20 Å/cycle at 300 °C [23]. For a similar $Gd(thd)_3$-based deposition process, Niinistö et al. reported self-limiting growth for Gd_2O_3 films in the range from 250 to 300 °C, with a GPC of $\approx$ 0.30 Å/cycle at 300 °C [24]. Our attempts at deposition of the same binary processes gave reproducible GPCs of 0.16 Å/cycle and 0.37 Å/cycle for the formation of CoO_x and Gd_2O_3, respectively, with no observed thickness gradients.

Based on these results, a series of (Gd, Co)-oxide films were deposited at 300 °C. The $Gd(thd)_3$: $Co(thd)_2$ pulsed ratio was varied systematically to identify the conditions required to obtain the desired deposited stoichiometry of $GdCoO_3$. We employed a super cycle approach with a general super cycle, given as:

$$n \times \{m \times [Gd(thd)_3 + O_3] + l \times [Co(thd)_2 + O_3]\}, \tag{1}$$

where n, m, and l were varied to achieve the desired cation ratio.

Figure 1 shows the deposited cation ratio for Gd (cat.% Gd) of the obtained film and the GPC as a function of the pulsed cation ratio (cat.% Gd) at 300 °C. The relative amount of deposited Gd increases from 2 to 51 cat.% Gd in the explored pulsed cation range of 33–67 cat.% Gd. The concentration of Gd in the deposited film consistently increases with increasing amounts of pulsed $Gd(thd)_3$, except for the plateau interval observed between 50 and 56 cat.% Gd pulsing ratio, where the Gd concentration in the product takes a constant value at around 32 cat.%. We note that the desired Gd:Co ratio of close to unity is obtained for 67 cat.% of pulsed Gd. The GPC of the deposited films at 300 °C increases smoothly with an increased fraction of Gd pulses, in accordance with the higher GPC of Gd_2O_3, see Figure 1. However, an excess of Gd pulses must be applied to achieve stoichiometric $GdCoO_3$. We do observe a small reduction in overall GPC (0.24 Å/cycle) as compared to a linear combination of the binary oxides [(0.37 + 0.16)/2 = 0.27 Å/cycle], possibly due to either inhibition of growth from $Gd(thd)_3$ on Co-O* surfaces or by increased growth from $Co(thd)_2$ on Gd-O* surfaces, or most likely a combination of both judging from the dependency of pulsed to deposited composition in Figure 1. This is an effect seen in several ALD processes, such as reported earlier by our group in the case of $LaAlO_3$ [25]. The GPCs obtained at 300 °C for (Gd, Co)-oxides are in good agreement with the results of Seim et al., who reported an average GPC of 0.35 Å/cycle for $LaCoO_3$ at 350 °C following a similar β-diketonate and ozone deposition process [21].

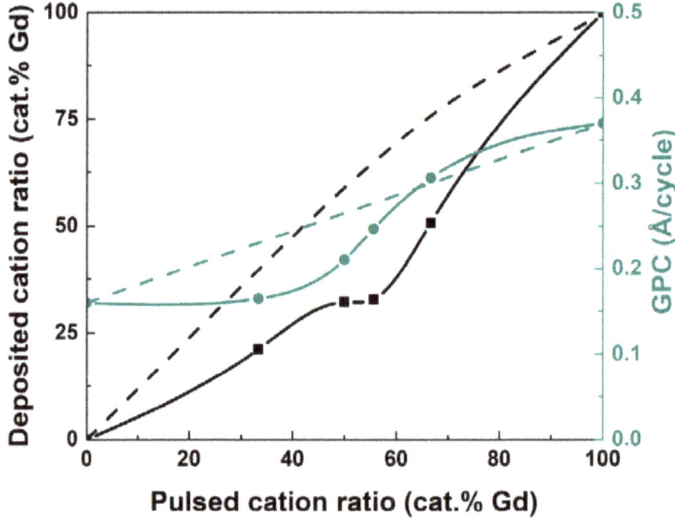

Figure 1. Deposited cation ratio (cat.% Gd, as measured by XRF) and GPC as a function of the pulsed cation ratio (cat.% Gd) for (Gd, Co)-oxide films deposited at 300 °C. The dotted lines refer to deposited cation ratio (cat.% Gd) and GPC as a function of the pulsed cation ratio (cat.% Gd) for Gd_2O_3 film.

3.2. Deposition of $Gd_{1-x}Ca_xCoO_3$

Based on the results obtained for the ternary (Gd, Co)-oxide system, the quaternary (Gd, Ca, Co)-oxide system was explored in an attempt to target products with the $Gd_{0.9}Ca_{0.1}CoO_3$ composition. ALD was carried out at 300 °C following an identical process as for ternary (Gd, Co)-oxide films, with the essential modification of substituting a number of $Gd(thd)_3$-pulses with $Ca(thd)_2$-pulses. The $[Gd(thd)_3 + Ca(thd)_2]$: $Co(thd)_2$ pulsed ratio was maintained at 2:1 in order to keep the deposited (Gd + Ca): Co atomic ratio close to unity. The Ca pulsed ratio, using $Ca(thd)_2$ as precursor, was varied from 3 to 7 cat.%. Figure 2a shows the deposited cation ratios and the GPC as a function of the Ca pulsed ratio (cat.%) for depositions at 300 °C. The Ca content in the films correlates fairly well with the relative amount of Ca pulses. In a few experiments deviating behavior was observed, which reflects the challenge of controlling the simultaneous growth of three different cation species [26]. However, quite a stable growth situation was obtained for the range around 4–5 cat.% Ca. The A-site (Gd + Ca): B-site (Co) stoichiometry was analyzed as function of the relative amount of Ca-pulses (Figure 2b). With the current pulsing strategy, the target (Gd + Ca): Co ratio close to unity is obtained for films deposited at a Ca pulsed ratio between 3.5 and 5 cat.%. The targeted composition $Gd_{0.9}Ca_{0.1}CoO_3$

is obtained for a Ca pulsed ratio of 4.5 cat.%, for which an equiatomic ratio is maintained between the perovskite A- and B-sites.

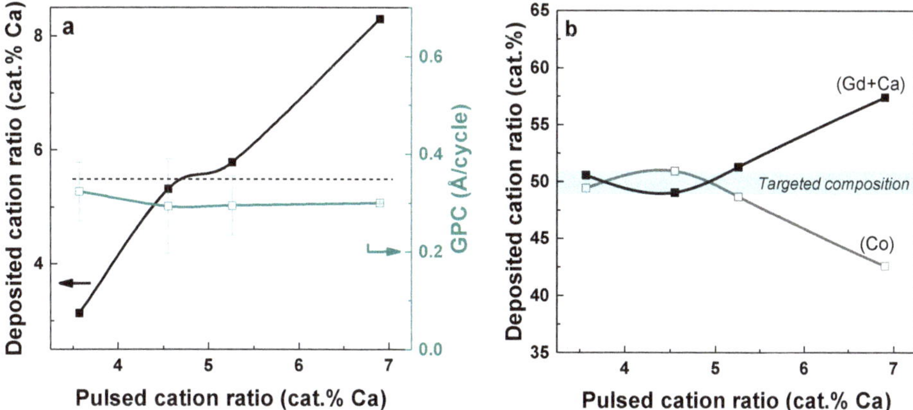

Figure 2. (**a**) Deposited cation ratio (cat.% Ca, as measured by XRF) and (**b**) GPC deposited cation ratio (cat. % (Gd + Ca) and Co, as measured by XRF) as a function of the pulsed cation ratio (cat.% Ca) for (Gd, Ca, Co) films deposited at 300 °C. The dotted line indicates the targeted Ca deposited concentration.

3.3. Characterization of GdCoO$_3$ and Gd$_{0.9}$Ca$_{0.1}$CoO$_3$ Thin Films

3.3.1. X-Ray Diffraction (XRD)

The as-prepared GdCoO$_3$ and Gd$_{0.9}$Ca$_{0.1}$CoO$_3$ films deposited at 300 °C are X-ray amorphous. Crystallization is achieved upon annealing at 650 °C for 30 min in air on LAO and YAP single crystals, resulting in preferential orientation depending on the substrate type and orientation. Figure 3a,b shows XRD patterns of post-annealed GdCoO$_3$ and Gd$_{0.9}$Ca$_{0.1}$CoO$_3$ films deposited on LAO(100)$_{pc}$. The diffractograms for the crystalline films on LAO(100)$_{pc}$ can be indexed as orthorhombic GdCoO$_3$ (*Pbnm*, SG# 62; Z = 4) with a preferred (010) growth orientation. The orthorhombically distorted GdCoO$_3$ perovskite relates to the ideal cubic perovskite structure (*Pm-3m*; Z = 1) as $a_o = \sqrt{2} \times a_c$, mboxemphb$_o$ = 2 × b_c, $c_o = \sqrt{2} \times c_c$ with dimensions a_o = 5.380 Å, b_o = 7.437 Å and c_o = 5.210 Å. The (rhombohedral) LAO substrate exhibits a pseudo cubic structure a_c = 3.79 Å (note $\sqrt{2} \times a_c$ = 5.36 Å). The growth of GdCoO$_3$-based perovskites onto LAO is favored in the (010) orientation as the a- and c-axis of the film match the diagonals of the cube faces of the substrate. In this configuration, GdCoO$_3$ will experience a lattice expansion of 2.5% in order to match the diagonal by diagonal area of the LAO substrate (A$_{LAO}$ = a_c^2 = 28.73 Å^2 and A$_{GCO}$ = $a_o \times c_o$ = 28.02 Å^2). The position of the (020) and (040) reflections indicate that $b_{GCO\|LAO(100)}$ = 7.42 Å (strain−0.2%), which indicates a small compression compared to the theoretical orthorhombic structure. This is in good agreement with the expected expansion in a. We used Scherrer's formula on the well-defined GdCoO$_3$ (040) reflection (Supporting Figure S1) to estimate a crystallite size of 24.8 nm, which indicates that the crystallites traverse from the substrate to the film surface. A higher degree of crystallinity is observed for GdCoO$_3$, which exhibits sharper and more intense (020) and (040) reflections than Gd$_{0.9}$Ca$_{0.1}$CoO$_3$. This is in good agreement with Bretos et al., who reported a slower crystallization process for Ca-substituted perovskites [27].

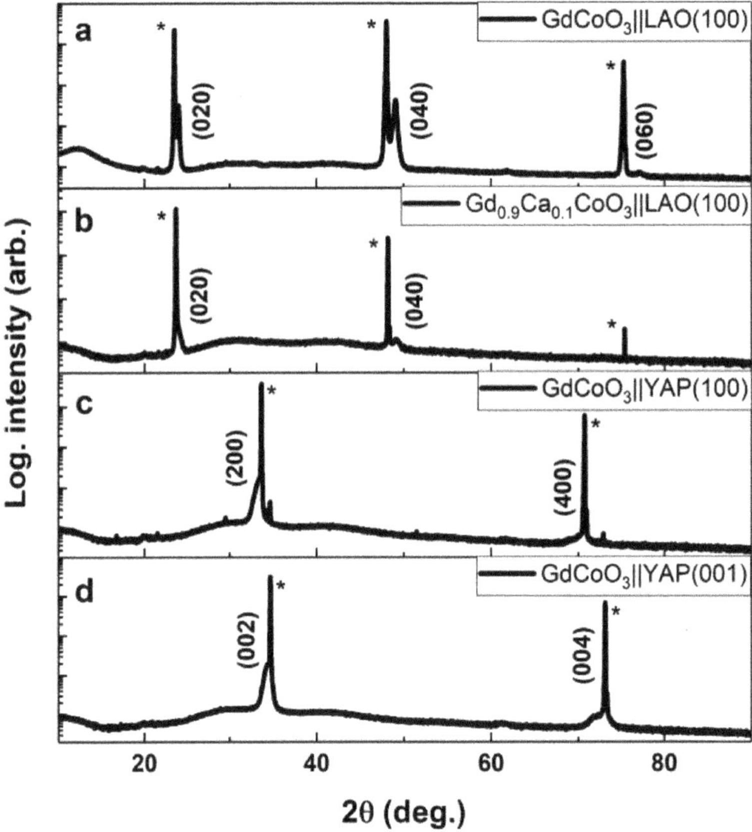

Figure 3. XRD patterns of (**a**) 30 nm GdCoO$_3$ and (**b**) 30 nm Gd$_{0.9}$Ca$_{0.1}$CoO$_3$ films grown on LAO(100), (**c**) 30 nm GdCoO$_3$ grown on YAP(100) and (**d**) 30 nm GdCoO$_3$ grown on YAP(001), post-annealed for 30 min at 650 °C. Bragg reflections originating from the substrate are marked with a star; film reflections are marked with their designated plane of reflection.

Figure 3c,d shows the measured XRD patterns from crystalline GdCoO$_3$ films deposited on YAP (100) and YAP (001), respectively, after post annealing at 650 °C in air for 30 min. Both YAP and GdCoO$_3$ are orthorhombic perovskites and exhibit quite similar unit cell dimensions; for YAP, a_{YAP} = 5.330 Å, b_{YAP} = 7.375 Å and c_{YAP} = 5.180 Å. GdCoO$_3$ deposited on YAP(001) grows with a preferred (001) orientation, whereas GdCoO$_3$ deposited on YAP(100) exhibits a preferential (100) growth orientation. Thus, the preferential (001) growth orientation of GdCoO$_3$ onto an oriented YAP(001) substrate is favored due to a minimized lattice compressive stress of 1.7% in this configuration ($V_{YAP(001)} = a_{YAP} \times b_{YAP}$ = 39.31 Å^2 and $V_{GCO} = a_o \times b_o$ = 39.97 Å^2). On the other hand, a (100) orientation of the substrate results in the growth of GdCoO$_3$ in a preferential (100) orientation with a lattice compression of 1.2% ($V_{YAP(100)} = b_{YAP} \times c_{YAP}$ = 38.20 Å^2 and $V_{GCO} = b_o \times c_o$ = 38.64 Å^2). The close lattice match means that the film reflections are observed as a broadening of the substrate peaks, making it difficult to analyze the diffraction in terms of crystallite size or strain. The Gd$_{0.9}$Ca$_{0.1}$CoO$_3$ thin films deposited on YAP substrates exhibited too poor a crystallinity, even after annealing, to be properly indexed (see Supplementary Figure S2).

By use of this appropriate selection of substrates, we have demonstrated the preferred crystallization along all three crystallographic axes. The effect of surface structure on catalytic activity is well known, so the ability to select the growth orientation of crystalline films may be of high importance.

The physical properties of gadolinium cobaltites depend, inter alia, on the temperature and cation substitutions, type, and concentration, which in turn may have a profound effect on the catalytic performance. For instance, an expanded lattice triggered by the substrate may stabilize the high-spin Co(III) configuration at temperatures lower than 800 K, i.e., the transition temperature for bulk GdCoO₃ [28]. Such scenarios are interesting from an ALD perspective, since key physical and chemical performance properties may be tuned by the choice of appropriate lattice-matching substrates.

3.3.2. XPS

Detailed XPS spectra close to the Co 2p binding energies were collected to identify the chemical state of cobalt in the films (Figure 4). Previous reports of ALD-grown cobalt oxide using β-diketonates and ozone indicated a mixed $2^+/3^+$ valence. The presence of Co^{2+} could indicate detrimental inclusions of Co_3O_4 in the grown films. Co^{2+} can be identified by an intense shake-up satellite feature at around 786 eV, whereas the Co^{3+} satellite is shifted towards 790 eV. Currently, we only observed Co^{3+} satellite features, indicating that the films are dominated by $GdCoO_3$. Based on the data and the complexity of Co XPS, however, we cannot rule out that some Co^{2+} is present in the films. Survey spectra and detailed scans of O 1 s and C 1 s can be found in the Supplementary Materials (Figure S3).

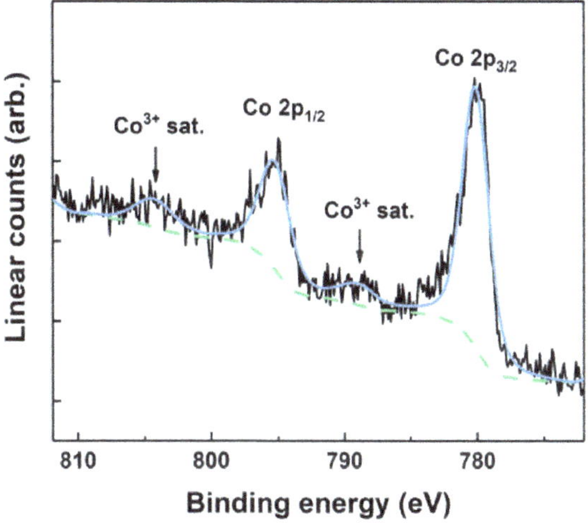

Figure 4. Co 2p XPS of 30 nm $GdCoO_3$ thin films on $LaAlO_3$ (100), with no observation of Co^{2+}. The black line is the recorded data, the green line is the background, and the light blue line is the total fit.

3.3.3. Deposition on a High-Aspect-Ratio Substrate

The ability to deposit catalytically active complex oxides on high aspect ratios is of high importance. This can, e.g., enable the coating of mesoporous γ-alumina, and thereby provide catalysts with a significantly enhanced surface area compared to nanoparticles (10–100 nm) obtained from wet chemical synthesis and/or ball milling. The conformality of the two gadolinium cobaltite-based films was investigated by applying the presented deposition process onto high-aspect-ratio substrates. Figure 5 shows cross section images of a $GdCoO_3$ film deposited on a high-aspect-ratio trench Si wafer. The film is conformally deposited on all surfaces of the substrate. As shown in Figure 5b, the bottom of the trench is characterized by the presence of agglomerates, possibly resulting from turbulence during growth and/or the preparation of cross section SEM samples.

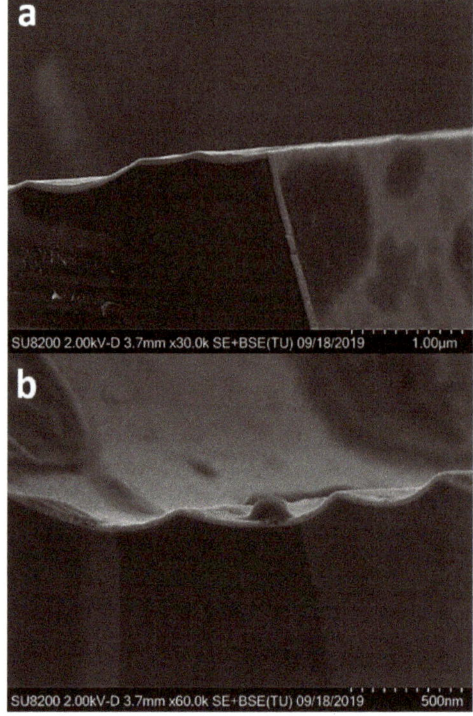

Figure 5. SEM cross-sectional images of $GdCoO_3$ deposited on a trench Si wafer: (**a**) top trench view; (**b**) bottom trench view.

4. Conclusions

We have developed an ALD process for crystalline and homogeneous Gd-Ca-Co-O thin films on three different substrates, which provides a route towards coatings with potential application within catalysis. The good crystallinity of the obtained films gives insight into the crystal structure of the product and orientation of crystallization, which is essential since catalytic performance depends on key parameters connected with the structure, chemistry, and electronic states of exposed surfaces. These features are shown to be tuned by appropriate choices of ALD precursors, substrates, and deposition/annealing conditions. The proof of concept of depositing conformal Gd-Ca-Co-O films on high-aspect-ratio substrates is important, since practical applications in catalysis would require high surface areas.

The gadolinium cobaltite-based catalysts represent a particularly interesting system, not only with respect to catalysis, but also to physical properties. For instance, an expanded lattice triggered by the substrate may possibly stabilize a high-spin Co(III) state. Such scenarios suggest that ALD can be used to tune resulting properties by means of appropriate lattice-matching substrates.

Supplementary Materials: The following are available online at http://www.mdpi.com/1996-1944/13/1/24/s1: Figure S1: XRD patterns of 30 nm $Gd_{0.9}Ca_{0.1}CoO_3$ films grown on (a) YAP(100) and (b) YAP(001), post-annealed for 30 minutes at 650 °C. Figure S2: XRD pattern of the $GdCoO_3$ (040) reflection of as deposited (black) and annealed (green) 30 nm films grown on LAO $(100)_{pc}$, used for Scherrer analysis of crystallite size. Figure S3: (a) XPS of C 1s, (b) XPS of O 1s and (c) Survey spectra showing identification of Gd, Co, O and carbon species.

Author Contributions: Investigation, M.D. and H.H.S.; formal analysis, M.D. and H.H.S.; methodology, H.H.S. and O.N.; project administration, O.N., A.O.S. and H.F.; writing—original draft preparation, M.D.; writing—review and editing, M.D., H.H.S., O.N., A.O.S. and H.F.; supervision, A.O.S. and H.F. All authors have read and agreed to the published version of the manuscript.

Funding: This project received financial support from the Research Council of Norway via the ASCAT project (contract no. 247753).

Acknowledgments: The authors acknowledge the Department of Geology at the University of Oslo for access to XRF instrumentation. Henrik H. Sønsteby acknowledges the Research Council of Norway for funding via the RIDSEM project (contract no. 272253). In addition, they thank Jon Einar Bratvold for his assistance with the ALD reactor, Kristian Weibye for recording XPS spectra, and Martin Jensen for recording SEM images.

Conflicts of Interest: The authors declare no conflict of interest.

References

1. Royer, S.; Duprez, D.; Can, F.; Courtois, X.; Batiot-Dupeyrat, C.; Laassiri, S.; Alamdari, H. Perovskites as Substitutes of Noble Metals for Heterogeneous Catalysis: Dream or Reality. *Chem. Rev.* **2014**, *114*, 10292–10368. [CrossRef] [PubMed]

2. Pena, M.A.; Fierro, J.L.G. Chemical Structures and Performance of Perovskite Oxides. *Chem. Rev.* **2011**, *101*, 1981–2018. [CrossRef] [PubMed]

3. Kim, C.H.; Qi, G.; Dahlberg, K.; Li, W. Strontium-doped perovskites rival platinum catalysts for treating NOx in simulated diesel exhaust. *Science* **2010**, *327*, 1624–1627. [CrossRef] [PubMed]

4. Fleming, P.; Farrell, R.A.; Holmes, J.D.; Morris, M.A. The Rapid Formation of La(OH)$_3$ from La$_2$O$_3$ Powders on Exposureto Water Vapor. *J. Am. Ceram. Soc.* **2010**, *93*, 1187–1194. [CrossRef]

5. Waller, D.; Grønvold, M.S.; Sahli, N. *Ammonia Oxidation Catalyst for the Production of Nitric Acid Based on Yttrium-Gadolinium Ortho Cobaltates*; World Intellectual Property Organization: New York, NY, USA, 2017.

6. Duparc, M.; et al. 2019; Unpublished manuscript.

7. Nagao, M.; Hamano, H.; Hirata, K. Hydration Process of Rare-Earth Sesquioxides Having Different Crystal Structures. *Langmuir* **2003**, *19*, 9201–9209. [CrossRef]

8. Krishnamurthy, N.; Gupta, C.K. *Extractive Metallurgy of Rare Earths*; CRC Press: Boca Raton, FL, USA, 2004.

9. Marichy, C.; Bechalany, M.; Pinna, N. Atomic Layer Deposition of Nanostructured Materials for Energy and Environmental Applications. *Adv. Mater.* **2012**, *24*, 1017–1032. [CrossRef]

10. O'Neill, B. Catalyst Design with Atomic Layer Deposition. *ACS Catal.* **2015**, *5*, 1804–1825. [CrossRef]

11. Camacho-Bunquin, J.; Shou, H.; Aich, P.; Beaulieu, D.R.; Klotzsch, H.; Bachman, S.; Marshall, C.L.; Hock, A.; Stair, P. Catalyst synthesis and evaluation using an integrated atomic layer deposition synthesis–catalysis testing tool. *Rev. Sci. Instrum.* **2015**, *86*, 84–103. [CrossRef]

12. Suntola, T. Atomic layer epitaxy. *Mater. Sci. Rep.* **1989**, *4*, 261–312. [CrossRef]

13. Leskelä, M.; Ritala, M. Atomic layer deposition (ALD): From precursors to thin film structures. *Thin Solid Films* **2002**, *409*, 138–146. [CrossRef]

14. George, S.M. Atomic layer deposition: An overview. *Chem. Rev.* **2009**, *110*, 111–131. [CrossRef] [PubMed]

15. Onn, T.M.; Dai, S.; Chen, J.; Pan, X.; Graham, G.W.; Gorte, R.J. High-Surface Area Ceria-Zirconia Films Prepared by Atomic Layer Deposition. *Catal. Lett.* **2017**, *147*, 1464–1470. [CrossRef]

16. Onn, T.M.; Monai, M.; Dai, S.; Arroyo-Ramirez, L.; Zhang, S.; Pan, X.; Graham, G.W.; Fornasiero, P.; Gorte, R.J. High-surface-area, iron-oxide films prepared by atomic layer deposition on γ-Al$_2$O$_3$. *Appl. Catal. A Gen.* **2017**, *534*, 70–77. [CrossRef]

17. Miikkulainen, V.; Leskelä, M.; Ritala, M.; Puurunen, R.L. Crystallinity of inorganic films grown by atomic layer deposition: Overview and general trends. *J. Appl. Phys.* **2013**, *113*, 2. [CrossRef]

18. Sønsteby, H.H.; Fjellvåg, H.; Nilsen, O. Functional Perovskites by Atomic Layer Deposition—An Overview. *Adv. Mater. Interfaces* **2017**, *4*, 1600903. [CrossRef]

19. Johnson, R.W.; Hultqvist, A.; Bent, S.F. A brief review of atomic layer deposition: From fundamentals to applications. *Mater. Today* **2014**, *17*, 236–246. [CrossRef]

20. Ahvenniemi, E.; Matvejeff, M.; Karppinen, M. Atomic layer deposition of quaternary oxide (La,Sr)CoO$_{3-\delta}$ thin films. *Dalton Trans.* **2015**, *44*, 8001–8006. [CrossRef]

21. Seim, H.; Nieminen, M.; Niinistö, L.; Fjellvåg, H.; Johansson, L.S. Growth of LaCoO$_3$ thin films from β-diketonate precursors. *Appl. Surf. Sci.* **1997**, *112*, 243–250. [CrossRef]

22. Sønsteby, H.H.; Bratvold, J.E.; Weibyc, K.; Fjellvåg, H.; Nilsen, O. Phase Control in Thin Films of Layered Cuprates. *Chem. Mater.* **2018**, *30*, 1095–1101. [CrossRef]

23. Klepper, K.B.; Nilsen, O.; Fjellvåg, H. Growth of thin films of Co_3O_4 by atomic layer deposition. *Thin Solid Fims* **2007**, *515*, 7772–7781. [CrossRef]

24. Niinistö, J.; Petrova, N.; Putkonen, M.; Niinistö, L.; Arstila, K.; Sajavaara, T. Gadolinium oxide thin films by atomic layer deposition. *J. Cryst. Growth* **2005**, *285*, 191–200. [CrossRef]

25. Sønsteby, H.H.; Østreng, E.; Fjellvåg, H.; Nilsen, O. Deposition and x-ray characterization of epitaxial thin films of $LaAlO_3$. *Thin Solid Films* **2014**, *550*, 90–94. [CrossRef]

26. Leskelä, M.; Ritala, M.; Nilsen, O. Novel materials by atomic layer deposition and molecular layer deposition. *MRS Bull.* **2011**, *36*, 877–884. [CrossRef]

27. Bretos, I.; Ricote, J.; Jiménez, R.; Mendiola, J.; Jiménez Riobóo, R.J.; Calzada, M.L. Crystallisation of $Pb_{1-x}Ca_xTiO_3$ ferroelectric thin films as a function of the Ca^{2+} content. *J. Eur. Ceram. Soc.* **2005**, *25*, 2325–2329. [CrossRef]

28. Orlov, Y.S.; Solovyov, L.A.; Dudnikov, V.A.; Fedorov, A.S.; Kuzubov, A.A.; Kazak, N.V.; Voronov, V.N.; Vereshchagin, S.N.; Shishkina, N.N.; Perov, N.S.; et al. Structural properties and high-temperature spin and electronic transitions in $GdCoO_3$: Experiment and theory. *Phys. Rev. B* **2013**, *81*, 235105. [CrossRef]

Article

Influence of Carrier Gases on the Quality of Epitaxial Corundum-Structured α-Ga$_2$O$_3$ Films Grown by Mist Chemical Vapor Deposition Method

Yu Xu, Chunfu Zhang *, Yaolin Cheng, Zhe Li, Ya'nan Cheng, Qian Feng, Dazheng Chen, Jincheng Zhang and Yue Hao

Wide Bandgap Semiconductor Technology Disciplines State Key Laboratory, School of Microelectronics, Xidian University, Xi'an 710071, China; xuyuxidian@163.com (Y.X.); chengyaolin96@163.com (Y.C.); zhe_li1024@163.com (Z.L.); yanancheng@stu.xidian.edu.cn (Y.C.); qfeng@mail.xidian.edu.cn (Q.F.); dzchen@xidian.edu.cn (D.C.); jchzhang@xidian.edu.cn (J.Z.); yhao@xidian.edu.cn (Y.H.)
* Correspondence: cfzhang@xidian.edu.cn

Received: 22 August 2019; Accepted: 4 November 2019; Published: 7 November 2019

Abstract: This report systematically investigates the influence of different carrier gases (O$_2$, N$_2$, and air) on the growth of gallium oxide (Ga$_2$O$_3$) thin films on c-plane sapphire substrates by using the mist-CVD method. Although XRD and Raman measurements show that the pure corundum-structured α-Ga$_2$O$_3$ with single (0006) plane orientation was successfully obtained for all three different carrier gases, the crystal quality could be greatly affected by the carrier gas. When O$_2$ is used as the carrier gas, the smallest full-width at half maximum (FWHM), the very sharp absorption cutoff edge, the perfect lattice structure, the highest growth rate, and the smooth surface can be obtained for the epitaxial α-Ga$_2$O$_3$ film as demonstrated by XRD, UV-VIS, TEM, AFM (Atomic Force Microscope), and SEM measurements. It is proposed that the oxygen content in carrier gas should be responsible for all of these results. XPS (X-ray photoelectron spectroscopy) analysis also confirms that more oxygen elements can be included in epitaxial film when O$_2$ is used as the carrier gas and thus help improve the crystal quality. The proper carrier gas is essential for the high quality α-Ga$_2$O$_3$ growth.

Keywords: wide-bandgap semiconductor; α-Ga$_2$O$_3$; mist chemical vapor deposition (mist-CVD); carrier gas; transparent semiconductor

---

1. Introduction

As an ultra-wide-bandgap semiconductor with the obvious advantages of stable physical chemistry, low dielectric constant, and high mechanical strength, gallium oxide (Ga$_2$O$_3$) is attracting increasing attention as a new promising competitor to III-nitrides and SiC for various applications in high-voltage and high-power electronics and ultraviolet optoelectronics [1]. Compared with other oxides such as ZnO (3.24 eV) and In$_2$O$_3$ (3.6 eV), Ga$_2$O$_3$ has a larger bandgap energy of approximately 5 eV, which means a shorter absorption cutoff wavelength and a much higher power application. There are a total of five different polytypes (α, β, ε, δ, and γ) for Ga$_2$O$_3$. Until now, the most studied polytype was β-Ga$_2$O$_3$ because it is easy to obtain bulk and film β-Ga$_2$O$_3$ materials by the conventional crystal growth or epitaxial growth techniques, such as edge-defined film-fed growth, float-zone method, Czochralski method, molecular beam epitaxy, and metal organic chemical vapor deposition. β-Ga$_2$O$_3$ has a bandgap of 4.8 eV and high Baliga's figures of merit (FOM) of 3000, which is obviously superior to GaN and SiC. However, β-Ga$_2$O$_3$ is not the best candidate in various phases for the power application considering the bandgap. Compared to β phase, corundum-structured α-Ga$_2$O$_3$, another important phase for Ga$_2$O$_3$, has a wider bandgap of around 5.3 eV which can result in a larger Baliga's FOM, in theory. Thus, α-Ga$_2$O$_3$ has great potential for application in power devices. A wider bandgap of about 5.3 eV means that the absorption cutoff wavelength can be shorter than 240 nm and then α-Ga_2O_3 is more suitable for ultraviolet optoelectronics. Although α-Ga_2O_3 has great application potential, the research of α-Ga_2O_3 still lags far behind β-Ga_2O_3 in large part because it is more difficult to obtain the high-quality α-Ga_2O_3 material than β-Ga_2O_3.

α-Ga_2O_3 is a metastable phase and the bulk material still cannot be obtained. There is no commercial α-Ga_2O_3 bulk substrate to date, so its homo-epitaxial growth is still difficult. Fortunately, the heterogeneous epitaxy provides an efficient way to obtain the α-Ga_2O_3 material. Corundum-structured gallium oxide belongs to the space group of R-3c with the lattice parameters a = b = 4.98 Å, c = 13.43 Å, $\alpha = \beta = 90°$, and $\gamma = 120°$ [2], and the lattice mismatches between α-Ga_2O_3 and α-Al_2O_3 (sapphire) are only 4.81% and 3.54% in the a- and c-axis directions. The same crystal structure and small lattice mismatch make it easy to grow the α-Ga_2O_3 material on the α-Al_2O_3 sapphire substrate. More importantly, the same crystal structure means that it is attractive from the viewpoint of fabricating alloys for α-Ga_2O_3 with other corundum-structured materials, such as α-Al_2O_3, Fe_2O_3, and Cr_2O_3 for bandgap and material engineering [3], which is another advantage of α-Ga_2O_3.

Recent studies have shown that the growth of crystalline α-Ga_2O_3 on an inexpensive sapphire substrate is an efficient way to obtain the α-Ga_2O_3 material. The key technology for the growth of α-Ga_2O_3 is ultrasonic mist chemical vapor deposition (mist-CVD) method [4–6]. In the growth of a metal oxide, water solutions of safe and inexpensive chemicals containing the metal, for example, acetate or acetylacetonate, have been used as the source. By atomizing the source solution ultrasonically, it turns into mist particles, which are then transferred by a carrier gas to a reaction chamber. In this way, metal elements are supplied without the use of organometallic sources. They react with an oxygen source, which may be water or oxygen gas. This offers sufficient overpressure of oxygen with respect to the metal source and prevents the formation of oxygen vacancies. Therefore, mist-CVD method is suitable for epitaxial α-Ga_2O_3 on sapphire substrate and can reduce the material cost [7,8]. Based on the grown α-Ga_2O_3, various applications of Metal Epitaxial-Semiconductor Field Effect Transistor [9], Schottky barrier diodes [10], and solar-blind photodetectors [11] have been demonstrated. For example, a high performance Schottky diode with a breakdown voltage over 1 kV and a small specific on-resistance of 2.5 mΩ·cm^2 has been achieved and a normally-off MOSFET has been shown based on α-Ga_2O_3 material grown by a mist-CVD system [9,10]. However, the present device performance is greatly inferior to the β-Ga_2O_3 counterparts and the main reason is still the poor α-Ga_2O_3 film quality. Thus, more attention is urgently required to improve the quality of α-Ga_2O_3 now and in the future. Recently, highly crystalline α-Ga_2O_3 thin films have been successfully grown at atmospheric pressure by mist-CVD on c-sapphire substrates, whose temperatures of 400–500 °C are reasonably low and the optimal growth conditions of solution concentration, growth temperature, carrier gas velocity, and film thickness have also been investigated [12,13]. However, there is still no systematical study about how the different carrier gas affects the film quality.

In this paper, we systematically investigate the influence of different carrier gases (O_2, N_2, and air) on the film quality for the growth of α-Ga_2O_3 on c-plane sapphire substrates by using the mist-CVD method. It is demonstrated that the crystallization quality will be different when the gallium source is carried by different gases. When N_2 and O_2 are used as the carrier gases, α-Ga_2O_3 achieves a relative smooth surface. When O_2 is the carrier gas, α-Ga_2O_3 achieves the smallest half-height width. The oxygen element in the carrier gas may be an important reason to prevent the generation of oxygen vacancies, thus influencing the quality of the thin films. The results provide constructive perspectives for the material quality improvement.

2. Materials and Methods

In the present experiment, we used gallium acetylacetonate as the gallium source, which was dissolved in deionized water. A small quantity of hydrochloric acid was added to dissolve gallium acetylacetonate completely. The concentration of the solution was adjusted to 0.05 M. By atomizing the source solution ultrasonically, it turned into mist particles (diameter of ~3 μm at an ultrasonic frequency of 2.4 MHz), which was then carried by air, N_2, and O_2, respectively, to the heated reaction chamber. In the chamber, the (0001) sapphire substrate was placed on a sample holder that was kept at 400 °C, because under this condition a high quality α-Ga_2O_3 can be obtained, as shown in Figure S1 in the supplemental information. The growth time was kept at 1h and the rate of the carrier gas was set to be 6 L/min.

The structural properties of Ga_2O_3 films were investigated using a range of complementary techniques. X-ray diffraction (XRD) patterns were obtained from an X-ray diffractometer (D8 Advance, Bruker, Karlsruhe, Germany). The transmittance was measured by a dual-beam 950 UV-VIS spectrometer. XPS measurements were performed by the Escalab 250Xi (Waltham, MA, USA) with a source of monochromatic Al-Ka (1486.6 eV). The film morphologies were characterized by a field emission scanning electron microscope (SEM JSM-7800F, Tokyo, Japan), atomic force microscopy (AFM) (Agilent 5500, Palo Alto, Santa Clara, CA, USA), and high-resolution transmission electron microscopy (TEM) (Tecnai G2 F20 S-Twin, Hillsboro, OR, USA). The Raman spectra were measured using a confocal Jobin Yvon LavRam HR800 micro-Raman spectrometer (Edison, NJ, USA) with a charge-coupled device (CCD) detector.

3. Results

Figure 1 shows the XRD spectra of the samples with air, N_2, and O_2 as the carrier gases. The spectra are dominated by the diffraction peaks at 40.26° and 41.66°, which correspond to the (0006) planes of α-Ga_2O_3 epilayer and sapphire substrate, respectively. No other peaks are found, which shows that all of the Ga_2O_3 films show the obvious pure alpha phase. These XRD spectra show that the α-Ga_2O_3 films had a preferential c-axis orientation along the c-axis of the sapphire substrate and the calculated lattice constant along the c-axis is 1.34 nm. Paying attention to the sample grown with O_2 as the carrier gas, the full-width at half maximum (FWHM) of the ω scan rocking curve is as small as 72 arcsec, indicating a high quality α-Ga_2O_3. However, for the samples grown with air and N_2 as the carrier gases, FWHM of the ω scan rocking curves are 88.6 arcsec and 86.4 arcsec, respectively, indicating a relatively inferior crystal quality. Considering the different oxygen content in the carrier gases, it is supposed that Ga_2O_3 film grown with O_2 as the carrier gas has less defects, such as oxygen vacancies (V_O), which may be the key to improving the quality of crystallization.

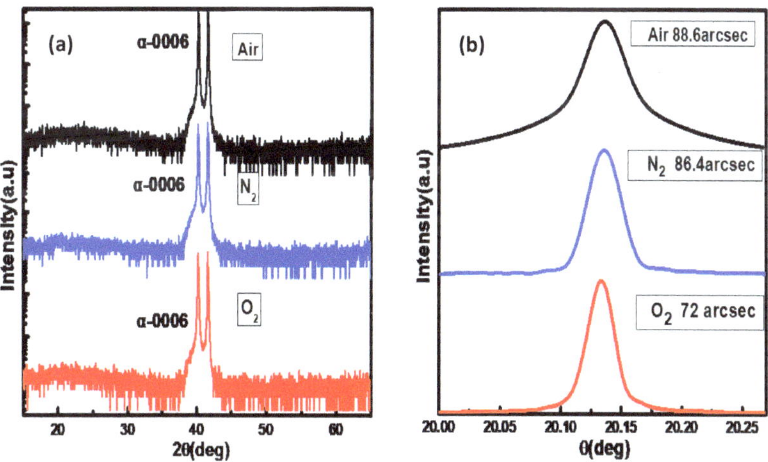

Figure 1. (a) X-ray diffraction 2θ/θ scan spectra for Ga_2O_3 films grown with different carrier gases. (b) The corresponding (0006) XRD diffraction rocking curves. The full-width at half maximum of Ga_2O_3 films grown with air, N_2, and O_2 as the carrier gases are 88.6 arcsec, 86.4 arcsec, and 72 arcsec, respectively.

Raman spectra of the epilayers grown with air, N_2, and O_2 as the carrier gases were measured between 100 cm^{-1} and 800 cm^{-1} at room temperature to confirm the crystalline quality of the deposited films, and the results are presented in Figure 2. A 514 nm laser was used as the excitation source and the laser beam was focused by a microscope lens system (×50 ulwd) yielding a spot size of 1 μm during the Raman measurement. The Raman peaks at 418 cm^{-1} and 749 cm^{-1} belong to the sapphire substrate [14]. The Raman peaks located at 431.3 cm^{-1}, 577 cm^{-1}, and 692 cm^{-1} are the Raman-allowed vibrational modes of E_g, A_{1g}, and E_g for α-Ga_2O_3, respectively, and are consistent with the theoretical calculations [15]. The high-frequency A_{1g} mode at 577 cm^{-1} mainly involves the vibration of oxygen atoms perpendicular to the c-axis. The linewidth of the peak A_{1g} is as narrow as 3.9 cm^{-1} for the epilayer grown with O_2 as the carrier gas, strongly suggesting the high crystallinity of the epilayer. By comparing the epilayers grown with different carrier gases, the displacement and intensity of the Raman peaks do not change obviously, demonstrating that the stress in the films is mainly determined by other factors, such as lattice mismatch or growth temperature, instead of the carrier gases. It could also be observed that except the Raman peaks of α-Ga_2O_3, no other peaks are observed in the Raman spectra, which indicates that all of the epilayers are pure α-Ga_2O_3 without other phases and confirm the conclusion from the XRD measurements.

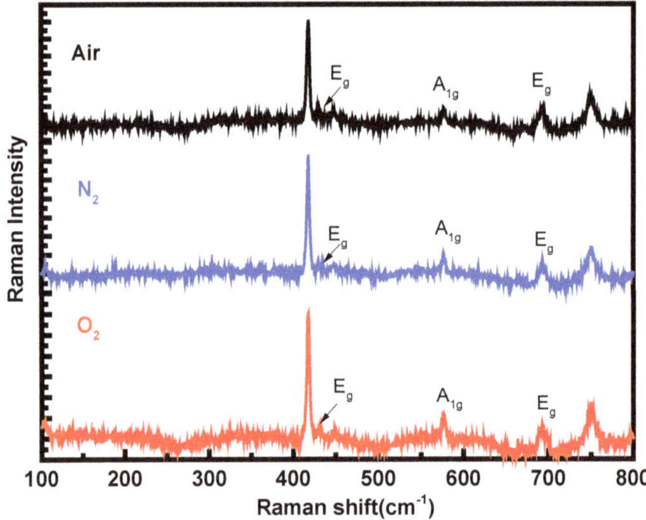

Figure 2. Raman spectra of the α-Ga_2O_3 epilayers grown with air, N_2, and O_2 as the carrier gases.

The variation of optical transmittance spectra (200–800 nm) were performed on α-Ga_2O_3 films as shown in Figure 3. All the samples exhibited a transmittance higher than 80% in the visible to near-UV regions. The relationship between the absorption coefficient α and the optical bandgap (E_g) is $\alpha h\nu = A(h\nu - Eg)^{1/2}$, where A is the material-dependent constant, h is the Planck's constant, and ν is the frequency of the incident light [16]. The optical band gap can be evaluated from the $(\alpha h\nu)^2$ versus photon energy ($h\nu$) graph by linear extrapolations to zero absorption coefficient [17]. The inset shows the plot of $(\alpha h\nu)^2$ as function of photon-energy $h\nu$. The bandgaps of the obtained materials remained at 5.1–5.3 eV, which is obviously larger than that of β-Ga_2O_3. The larger bandgap than that of β-Ga_2O_3 is obviously attributed to the crystal structure of α-Ga_2O_3 being different from β-Ga_2O_3. We note that the α-Ga_2O_3 sample grown with O_2 as the carrier gas is dropped more abruptly than those of air and N_2 in ultraviolet region. This phenomenon may be caused by less defect in α-Ga_2O_3 samples carried by O_2. During the film growth, extra oxygen will ensure the ideal ratio of O to Ga atomic, which results in higher crystalline quality and leads to its bandgap being close to the ideal value of

5.3 eV. The higher crystalline quality for the sample with O_2 as the carrier gas is consistent with the XRD measurement results.

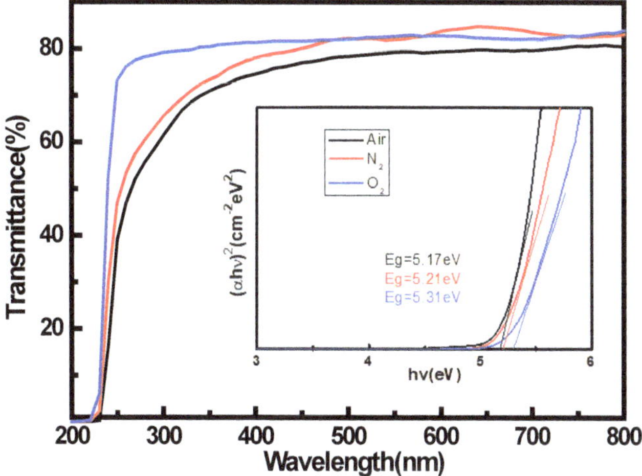

Figure 3. Optical transmission spectra and $(\alpha hv)^2$-hv plots of α-Ga_2O_3 films.

Figure 4a shows the electron diffraction patterns of the Ga_2O_3/Al_2O_3 interface for the sample with O_2 as the carrier gas. The diffraction patterns of both the Ga_2O_3 film and the Al_2O_3 substrate are rectangular, corresponding to the corundum-structure. All of the diffraction spots of α-Ga_2O_3 are situated almost in the α-Al_2O_3 spots. The cross-sectional HR-TEM image at the α-Ga_2O_3/α-Al_2O_3 interface is shown in Figure 4b. As a result of the in-plane strain, the α-Ga_2O_3/α-Al_2O_3 interface is unclearly identified. At the α-Ga_2O_3/α-Al_2O_3 interface, we can observe a dark area and this is induced by the in-plane compressive strain in the α-Ga_2O_3 layer because the lattice constant of α-Ga_2O_3 is larger than that of α-Al_2O_3 substrate. The TEM images for the samples, with air and N_2 as the carrier gases in Figures S1 and S2 also demonstrate the similar electron diffraction patterns and obvious HR-TEM lattice structures, indicating that all of the samples achieve a pure α-Ga_2O_3 phase and confirm the conclusion from the XRD measurements again. By comparing the HR-TEM images for the three different samples, the samples with O_2 and air as the carrier gases show a more complete lattice structure than the sample with N_2 as the carrier gas, which shows that the adequate efficient oxygen content is essential for the high quality film growth and this result is the same with that from the XRD measurement. The α-Ga_2O_3 films deposited on sapphire substrates obtain the thicknesses of 619 nm, 318 nm, and 146 nm for the samples with O_2, air, and N_2 as the carrier gases, respectively. These thicknesses correspond to the different growth rates of 10.3 nm/min, 5.3 nm/min, and 2.4 nm/min as shown in Figure 4c. Variable-angle Spectral Ellipsometry (SE) measurements were performed at room temperature in ambient atmosphere with an electronically controlled rotating compensator and Glan Taylor polarizers (J. A. Woollam Co., Lincoln, NE, USA). The Cauchy model was used to fit the thickness of Ga_2O_3 thin film. Measurements were carried out at three different incidence angles of 55°, 65°, and 75° over the 193–1000 nm wavelength range. Considering the different oxygen contents in the carrier gases, it can be concluded that more oxygen in the carrier gas can promote the growth rate greatly, and at the same time it can also guarantee the crystal quality as demonstrated by the XRD, UV-VIS and TEM measurements.

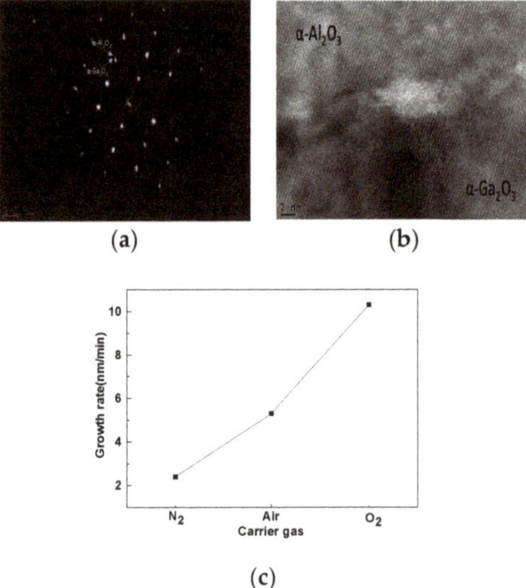

(a)   (b)

(c)

Figure 4. TEM images. (**a**) Diffraction spots of $\alpha\text{-Ga}_2\text{O}_3/\alpha\text{-Al}_2\text{O}_3$ for the sample with O_2 as the carrier gas, (**b**) cross-sectional $\alpha\text{-Ga}_2\text{O}_3/\alpha\text{-Al}_2\text{O}_3$ interface for the sample with O_2 as the carrier gas. (**c**) Growth rate of $\alpha\text{-Ga}_2\text{O}_3$ films grown by air, N_2 and O_2 carrier gas.

Figure 5 shows the AFM images for the $\alpha\text{-Ga}_2\text{O}_3$ films prepared by different carrier gases of air (a), N_2 (b), and O_2(c). The films have root-mean-square (RMS) surface roughness values of 6.6 nm, 1.16 nm and 2.18 nm, respectively, measured over an area of $5 \times 5~\mu\text{m}^2$. The surface morphology for the sample grown with air as the carrier gas is rough and composed of irregular stripes, while the surface morphologies of $\alpha\text{-Ga}_2\text{O}_3$ thin films grown with N_2 and O_2 as the carrier gases are much smoother. It is inferred that the complex components in air induced this rough surface, and the reason behind that requires more research in the future. On the contrary, the pure carrier gas will lead to a much smoother surface. Comparing the sample grown with O_2 as the carrier gas to the sample with N_2 as the carrier gas, although the growth rate is improved by about five times, the difference between their RMS values is relatively small. This means that the rapid growth rate with O_2 as the carrier gas will not degrade the surface morphology. To further investigate the microstructure, we measured SEM images of all $\alpha\text{-Ga}_2\text{O}_3$ films as shown in Figure 5d–f. As demonstrated in the enlarged pictures, it can be seen that the film grown with air as the carrier gas has indeed a rough surface, and the other samples show the relatively smooth surface, which confirm the results from the AFM measurement.

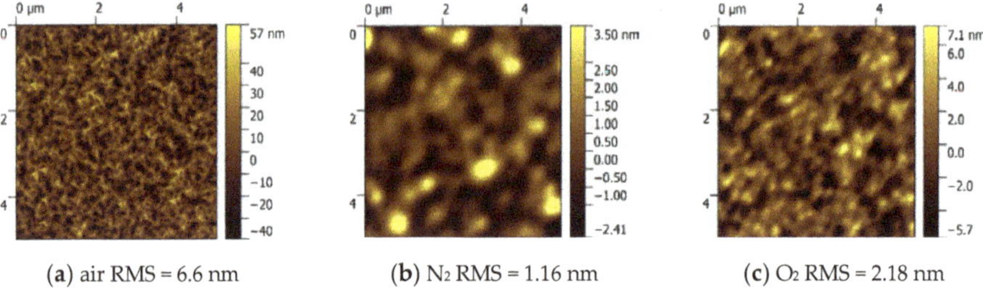

(**a**) air RMS = 6.6 nm   (**b**) N_2 RMS = 1.16 nm   (**c**) O_2 RMS = 2.18 nm

Figure 5. *Cont.*

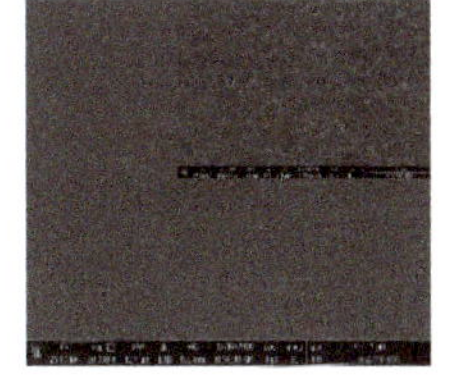

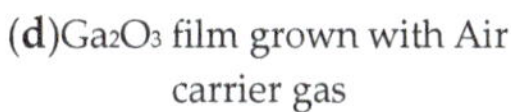

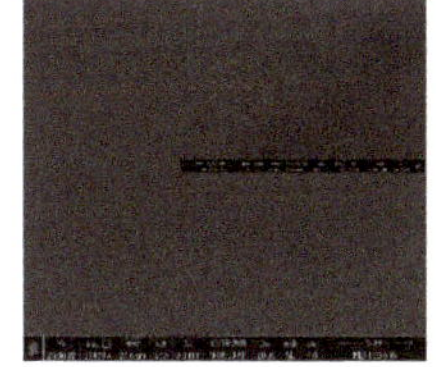

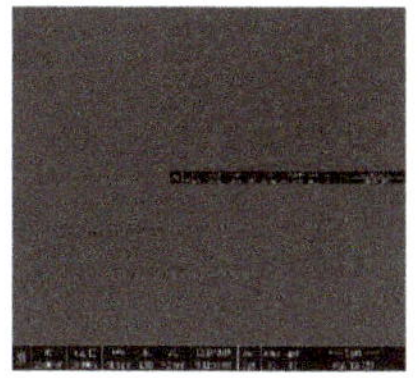

(**d**)Ga₂O₃ film grown with Air carrier gas   (**e**) Ga₂O₃ film grown with N₂ carrier gas   (**f**) Ga₂O₃ film grown with O₂ carrier gas

Figure 5. AFM surface images of Ga_2O_3 films grown with different carrier gases: (**a**) air (**b**) N_2, and (**c**) O_2. SEM pictures for the Ga_2O_3 films grown with different carrier gases: (**d**) air (**e**) N_2, and (**f**) O_2.

To assess the elemental composition of the α-Ga_2O_3 films, XPS measurements were conducted for the epitaxial films on sapphire. The result for the sample with O_2 as the carrier gas is presented in Figure 6a and only three elements (C, Ga, and O) are observed in the films. The position of the Ga $2p_{3/2}$ and Ga $3d$ binding energy peaks confirms the presence of Ga_2O_3 [18]. It can be observed from the spectra that the $Ga2p_{3/2}$ and $Ga2p_{1/2}$ signal peaks are located at 1118.5 eV and 1145.5 eV respectively, The energy difference between the two signal peaks is about 27 eV, which is consistent with the $Ga2p$ signal peak energy difference reported in the literature [19]. The binding energy position of $Ga3d$ signal is 21.05 eV, which is consistent with the results reported in the literature [20]. The result for the samples with air as the carrier gases is shown in Figures S3 and S4. The same peak positions appear in both samples and no significant peak shift and peak intensity changes are observed, indicating the same pure α-Ga_2O_3 and correspond to the XRD and Raman results. Figure 6b–d show the peaks of O $1s$ signal around 530.6 eV for the three samples. O $1s$ signal peaks do not show Gaussian symmetry so Gaussian fitting was carried out for the peaks. It is found that the strong O $1s$ signal peaks appear at 530.6 eV and 532.1 eV. The peak at 530.6 eV originated from the oxygen element in Ga–O bond, while the peak at 532.1 eV originated from the adsorption O on the sample surface [21]. Figure 5e–f show Ga $3d$ spectra. The chemical compositions of the surfaces were determined from the area of the Ga $3d$, O $1s$, and C $1s$ peaks taking into account the corresponding sensitivity factors [22]. The calculated atomic ratios of O to Ga for the samples grown with air, N_2, and O_2 as the carrier gases were 1.53, 1.51, and 1.56, respectively. It reveals that more oxygen present in the carrier gas can slightly increase the O content of the film, which confirms the guess in the XRD measurement.

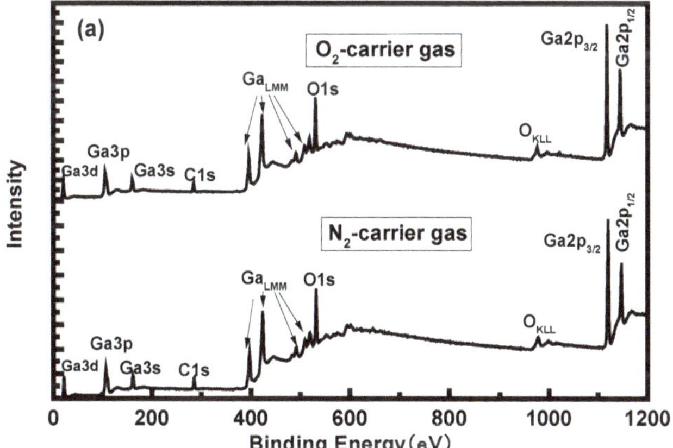

Figure 6. *Cont.*

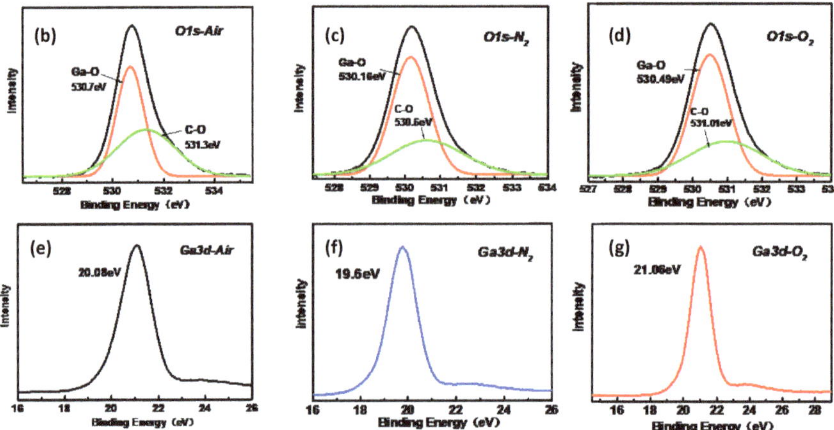

Figure 6. (**a**) X-ray photoelectron wide spectra for the α-Ga$_2$O$_3$ sample grown with O$_2$ and N$_2$ as the carrier gas. X-ray photoelectron spectra of O 1s peaks for the samples grown with air (**b**), N$_2$ (**c**), and O$_2$ (**d**) as the carrier gases. X-ray photoelectron spectra of Ga 3d peaks for the samples grown with air (**e**), N$_2$ (**f**), and O$_2$ (**g**) as the carrier gases.

We also compared the quality of the same thickness and growth rate of the α-Ga$_2$O$_3$ films grown by different carrier gases. As shown in Figure S2 in the supplemental information, the growth rate of the α-Ga$_2$O$_3$ films was set to 10 nm/min for all samples. In Figure S3 in the supplemental information, the thicknesses of all α-Ga$_2$O$_3$ samples were 500 nm. The optical transmittance spectra show that when O$_2$ is the carrier gas, the absorption edge is closer to 234 nm of α-Ga$_2$O$_3$ and the downward trend of the absorption edge is more obvious. From the AFM images, we assume that with the same growth rate and film thickness, the complex components in air induces rough surface and the pure carrier gas will lead to a much smoother surface. These results fully demonstrate that the partial pressure of oxygen in the carrier gas component can improve the quality of the film, consistent with the above results.

4. Conclusions

In this work, we systematically investigated the influence of different carrier gases (O$_2$, N$_2$, and air) on the grown film quality on c-plane sapphire substrates by using the mist-CVD method. Although XRD and Raman measurements show that the pure corundum-structured α-Ga$_2$O$_3$ was successfully obtained on the c-plane sapphire substrates with single (0006) plane orientation for all three different carrier gases, the crystal quality could be greatly affected by the carrier gas. When O$_2$ was used as the carrier gas, the sample showed the smallest FWHM of about 72 arcsec obtained from the XRD rocking curves. The UV-VIS measurement also showed that the sample grown with O$_2$ as the carrier gas showed a very sharp absorption cutoff edge. It is proposed that the oxygen content in carrier gas should be responsible for all of these results. More oxygen in the carrier gas can greatly increase growth rate, and at the same time guarantee the film quality. This is further verified by the smoother surface and rapid growth rate (10.3 nm/min) for O$_2$ as the carrier gas compared to air and N$_2$ as the carrier gases (5.3 nm/min and 2.4 nm/min respectively). XPS analysis confirmed that more oxygen element can be included in epitaxial film and thus help improve the crystal quality. The proper carrier gas is essential for high quality α-Ga$_2$O$_3$ growth.

Supplementary Materials: The following are available online at http://www.mdpi.com/1996-1944/12/22/3670/s1, Figure S1: TEM images of the sample grown with air as the carrier gas. (1) Cross-sectional α-Ga$_2$O$_3$/α-Al$_2$O$_3$ interface, (2) diffraction spots of α-Ga$_2$O$_3$/α-Al$_2$O$_3$, Figure S2: TEM images of the sample grown with N$_2$ as the carrier gas. (1) Cross-sectional α-Ga$_2$O$_3$/α-Al$_2$O$_3$ interface, (2) diffraction spots of α-Ga$_2$O$_3$/α-Al$_2$O$_3$, Figure S3:

X-ray photoelectron wide spectra for the α-Ga_2O_3 sample grown with air as the carrier gas, Figure S4: X-ray photoelectron wide spectra for the α-Ga_2O_3 sample grown with N_2 as the carrier gas.

Author Contributions: Conceptualization, C.Z.; methodology, C.Z. and Y.X.; validation, Y.C. (Yaolin Cheng), Z.L., Y.C. (Ya'nan Cheng); formal analysis, Y.X. and Y.C. (Yaolin Cheng); investigation, C.Z. and J.Z.; resources, Y.H.; data curation, D.C.; writing—original draft preparation, Y.X.; writing—review and editing, C.Z.; visualization, Q.F.; project administration, C.Z.; funding acquisition, Y.H.

Funding: This research was funded by National Key R&D Program of China, grant number 2018YFB0406504.

Conflicts of Interest: The authors declare no conflict of interest.

References

1. Higashiwaki, M.; Sasaki, K.; Kuramata, A.; Masui, T.; Yamakoshi, S. Development of gallimoxide power devices. *Phys. Status Solidi A* **2014**, *211*, 21–26. [CrossRef]
2. Kim, K.H.; Ha, M.T.; Kwon, Y.J.; Lee, H.; Jeong, S.M.; Bae, S.Y. Growth of 2-Inch α-Ga_2O_3 Epilayers via rear-flow-controlled mist chemical vapor deposition. *ECS J. Solid State Sci. Technol.* **2019**, *8*, Q3165–Q3170. [CrossRef]
3. Ito, H.; Kaneko, K.; Fujita, S. Growth and band gap control of corundum-structured α-$(AlGa)_2O_3$ thin films on sapphire by spray-assisted mist chemical vapor deposition. *J. Appl. Phys.* **2012**, *5*, 100207. [CrossRef]
4. Nakabayashi, Y.; Yamada, S.; Itoh, S.; Kawae, T. Influence of precursor concentration and growth time on the surface morphology and crystallinity of α-Ga_2O_3 thin films fabricated by mist chemical vapor deposition. *J. Ceram. Soc. Jpn.* **2018**, *126*, 925–930. [CrossRef]
5. Lee, S.D.; Akaiwa, K.; Fujita, S. Thermal stability of single crystalline alpha gallium oxide films on sapphire substrates. *Phys. Status Solidi-R.* **2013**, *10*, 1592–1595.
6. Shirahata, T.; Kawaharamura, T.; Fujita, S.; Orita, H. Transparent conductive zinc-oxide-based films grown at low temperature by mist chemical vapor deposition. *Thin Solid Film.* **2015**, *597*, 30–38. [CrossRef]
7. Shinohara, D.; Fujita, S. Heteroepitaxy of corundum-structured α-Ga_2O_3 thin films on α-Al_2O_3 substrates by ultrasonic mist chemical vapor deposition. *J. Appl. Phys.* **2008**, *47*, 7311–7313. [CrossRef]
8. Kaneko, K.; Nomura, T.; Kakeya, I.; Fujita, S. Fabrication of highly crystalline corundum-structured α-$(Ga1-xFex)_2O_3$ alloy thin films on sapphire substrates. *Appl. Phys. Express* **2009**, *2*, 075501. [CrossRef]
9. Dang, G.T.; Kawaharamura, T.; Furuta, M.; Allen, M.W. Mist-CVD Grown Sn-Doped\alpha-Ga_2O_3 MESFETs. *IEEE Trans. Electrons Dev.* **2015**, *62*, 3640–3644. [CrossRef]
10. Oda, M.; Tokuda, R.; Kambara, H.; Tanikawa, T.; Sasaki, T.; Hitora, T. Schottky barrier diodes of corundum-structured gallium oxide showing on-resistance of 0.1 mΩ·cm^2 grown by MIST EPITAXY®. *Appl. Phys. Express* **2016**, *9*, 021101. [CrossRef]
11. Guo, D.Y.; Zhao, X.L.; Zhi, Y.S.; Cui, W.; Huang, Y.Q.; An, Y.H.; Tang, W.H. Epitaxial growth and solar-blind photoelectric properties of corundum-structured α-Ga_2O_3 thin films. *Mater. Lett.* **2016**, *164*, 364–367. [CrossRef]
12. Ma, T.; Chen, X.; Ren, F.; Zhu, S.; Gu, S.; Zhang, R.; Ye, J. Heteroepitaxial growth of thick α-Ga_2O_3 film on sapphire (0001) by MIST-CVD technique. *J. Semicond.* **2019**, *40*, 012804. [CrossRef]
13. Kawaharamura, T.; Dang, G.T.; Furuta, M. Successful growth of conductive highly crystalline Sn-doped α-Ga_2O_3 thin films by fine-channel mist chemical vapor deposition. *Jpn. J. Appl. Phys.* **2012**, *51*, 0207.
14. Zhao, Y.; Frost, R.L. Raman spectroscopy and characterisation of α-gallium oxyhydroxide and β-gallium oxide nanorods. *J. Raman Spectrosc.* **2008**, *39*, 1494–1501. [CrossRef]
15. Cuscó, R.; Domènech-Amador, N.; Hatakeyama, T.; Yamaguchi, T.; Honda, T.; Artús, L. Lattice dynamics of a mist-chemical vapor deposition-grown corundum-like Ga_2O_3 single crystal. *J. Appl. Phys.* **2015**, *117*, 185706. [CrossRef]
16. Guo, D.; Wu, Z.; Li, P.; An, Y.; Liu, H.; Guo, X.; Tang, W. Fabrication of β-Ga_2O_3 thin films and solar-blind photodetectors by laser MBE technology. *Opt. Mater. Express* **2014**, *4*, 1067–1076. [CrossRef]
17. Ou, S.L.; Wuu, D.S.; Fu, Y.C.; Liu, S.P.; Horng, R.H.; Liu, L.; Feng, Z.C. Growth and etching characteristics of gallium oxide thin films by pulsed laser deposition. *Mater. Chem. Phys.* **2012**, *133*, 700–705. [CrossRef]
18. Thomas, S.R.; Adamopoulos, G.; Lin, Y.H.; Faber, H.; Sygellou, L.; Stratakis, E.; Anthopoulos, T.D. High electron mobility thin-film transistors based on Ga_2O_3 grown by atmospheric ultrasonic spray pyrolysis at low temperatures. *Appl. Phys. Lett.* **2014**, *105*, 092105. [CrossRef]

19. Kong, L.; Ma, J.; Luan, C.; Mi, W.; Lv, Y. Structural and optical properties of heteroepitaxial beta Ga_2O_3 films grown on MgO (100) substrates. *Thin Solid Film.* **2012**, *520*, 4270–4274. [CrossRef]
20. Dong, L.; Jia, R.; Xin, B.; Zhang, Y. Effects of post-annealing temperature and oxygen concentration during sputtering on the structural and optical properties of β-Ga_2O_3 films. *J. Vac. Sci. Technol. A* **2016**, *34*, 060602. [CrossRef]
21. Hueso, J.L.; Espinós, J.P.; Caballero, A.; Cotrino, J.; González-Elipe, A.R. XPS investigation of the reaction of carbon with NO, O_2, N_2 and H_2O plasmas. *Carbon* **2007**, *45*, 89–96. [CrossRef]
22. Ramana, C.V.; Rubio, E.J.; Barraza, C.D.; Miranda Gallardo, A.; McPeak, S.; Kotru, S.; Grant, J.T. Chemical bonding, optical constants, and electrical resistivity of sputter-deposited gallium oxide thin films. *J. Appl. Phys.* **2014**, *115*, 043508. [CrossRef]

materials

MDPI

Article

Nanoscale Piezoelectric Properties and Phase Separation in Pure and La-Doped BiFeO$_3$ Films Prepared by Sol–Gel Method

Alina V. Semchenko [1], Vitaly V. Sidsky [1], Igor Bdikin [2], Vladimir E. Gaishun [1], Svitlana Kopyl [3], Dmitry L. Kovalenko [1], Oleg Pakhomov [4], Sergei A. Khakhomov [1] and Andrei L. Kholkin [3,*]

[1] Faculty of Physics and Information Technology, Francisk Skorina Gomel State University, 246019 Gomel, Belarus; alina@gsu.by (A.V.S.); sidsky@gsu.by (V.V.S.); Vgaishun@gsu.by (V.E.G.); dkov@gsu.by (D.L.K.); khakh@gsu.by (S.A.K.)

[2] TEMA, Department of Mechanical Engineering, University of Aveiro, 3810-193 Aveiro, Portugal; bdikin@ua.pt

[3] Department of Physics, CICECO-Aveiro Institute of Materials, University of Aveiro, 3810-193 Aveiro, Portugal; svitlanakopyl@ua.pt

[4] Laboratory for "Electro-and Magnetocaloric Materials and Structures", ITMO University, 197101 St. Petersburg, Russia; oleg.cryogenics@gmail.com

* Correspondence: kholkin@ua.pt

Abstract: Pure BiFeO$_3$ (BFO) and doped Bi$_{0.9}$La$_{0.1}$FeO$_3$ (BLFO) thin films were prepared on Pt/TiO$_2$/SiO$_2$/Si substrates by a modified sol–gel technique using a separate hydrolysis procedure. The effects of final crystallization temperature and La doping on the phase structure, film morphology, and nanoscale piezoelectric properties were investigated. La doping and higher crystallization temperature lead to an increase in the grain size and preferred (102) texture of the films. Simultaneously, a decrease in the average effective piezoelectric coefficient (about 2 times in La-doped films) and an increase in the area of surface non-polar phase (up to 60%) are observed. Phase separation on the films' surface is attributed to either a second phase or to a non-polar perovskite phase at the surface. As compared with undoped BFO, La-doping leads to an increase in the average grain size and self-polarization that is important for future piezoelectric applications. It is shown that piezoelectric activity is directly related to the films' microstructructure, thus emphasizing the role of annealing conditions and La-doping that is frequently used to decrease the leakage current in BFO-based materials.

Keywords: bismuth ferrite; La-doping; piezoelectricity; sol–gel

Citation: Semchenko, A.V.; Sidsky, V.V.; Bdikin, I.; Gaishun, V.E.; Kopyl, S.; Kovalenko, D.L.; Pakhomov, O.; Khakhomov, S.A.; Kholkin, A.L. Nanoscale Piezoelectric Properties and Phase Separation in Pure and La-Doped BiFeO$_3$ Films Prepared by Sol–Gel Method. *Materials* **2021**, *14*, 1694. https://doi.org/10.3390/ma14071694

Academic Editor: Philippe Colomban

Received: 20 January 2021
Accepted: 25 March 2021
Published: 30 March 2021

Publisher's Note: MDPI stays neutral with regard to jurisdictional claims in published maps and institutional affiliations.

1. Introduction

Recently, there has been a rising interest in multiferroic materials, which demonstrate both magnetic and polarization order and resulting coupling between them in a single phase [1–7]. If an external electric field is applied, multiferroics can change their magnetic moment (converse magnetoelectric effect), and magnetic field can influence dielectric parameters such as electric polarization (direct magnetoelectric effect) or dielectric constant (magneto capacitance). As such, the possibility of manipulating the electric parameters via a magnetic field and may lead to new applications such as magnetic/electric memories, spintronics, magnetocapacitive transducers, magnetic field sensors, etc. Bismuth ferrite, BiFeO$_3$ (BFO), is one of the most studied multiferroics having perovskite structure (space group *R3c*) and exceptionally high ferroelectric and magnetic Curie points (830 °C and 402 °C, respectively). The last few years have witnessed a continuing interest in BFO due to the observation of the large conductivity of ferroelectric domain walls and the possibility to manipulate both magnetic and ferroelectric properties by appropriate doping. Weak magnetization of BFO due to antiferromagnetic G-type ordering restricts its application, and there is a chance for magnetization to be increased by doping. One of the biggest problems in BiFeO$_3$ films is their large leakage current, which strongly affects both ferroelectric and dielectric properties. One of the disadvantages of BiFeO$_3$ films is due to their high conductivity explained by different oxidation states of Fe ions, Fe^{3+} and Fe^{2+}, which result in the appearance of oxygen vacancies that are responsible for the hopping conduction [8,9]. Conductivity in $BiFeO_3$ films has to be decreased to improve their use in electronic applications. One of the efficient means is to dope them at different sites of the perovskite lattice. Doping of BFO has been extensively studied in the past: there were reports on Tb [10], La [11,12], Ce [13], Eu, Gd, Dy [14] doping for A-site and Ti [8], Cr [15], Zr [16], Mn [17] substitutions at B-site. Using these substitutions it was possible to stabilize the valence of iron with the simultaneous decrease of conductivity. In particular, La-doping was frequently used to create structurally metastable states with both enhanced piezoelectricity and reduced leakage [8,9]. These states, characterized by notable magnetic properties, may occur at the boundaries of nanosized grains and various interfaces in thin films. For example, the bottom electrode's resistance was found to affect ferroelectric switching kinetics [1]. Magnetic field effect on polarization switching in BFO films was studied in [2]. Magnetic ordering as a function of mechanical strain was found to be important in $BiFeO_3$ films [4]. However, no detailed nanoscale measurements of piezoelectric activity in doped $BiFeO_3$ films have not been reported yet.

The sol–gel method can be applied to process a variety of materials, including thin films with controlled functional properties [18]. Most of the materials prepared by sol–gel are have improved properties and broadly used in modern technologies [19–21]. During the synthesis of multicomponent sol–gel materials, the difficulties associated with different reaction rates for various elements similar to those described in [22–24] must be solved. In this work, a modified sol–gel process, in which the initial components are dissolved separately to form homogeneous organic solutions, followed by hydrolysis and polycondensation of the reaction products, leading to the formation of sol, and then to the final colloidal phase was used. The distinctive feature of the technique used in this article is separate hydrolysis procedures for metal (Bi, Fe and La) precursors in order to reach higher chemical homogeneity and to reduce the final crystallization temperature.

2. Materials and Methods

$BiFeO_3$ (BFO) thin films were synthesized using the standard sol–gel technique. $Bi(NO_3)_3 \cdot 5H_2O$) and $Fe(NO_3)_3 \cdot 5H_2O$ were used as raw reactants with ethylene glycol, dimethyl formamide, and citric were the solvents. All the chemical reagents used were obtained from MilliporeSigma (analytical grade, Munich, Germany). The original feature of the method is the use of separate hydrolysis, that is, each of the metal salts were separately dissolved in the mixture of solvents, kept for 24 h, then the sols were mixed and additionally kept for 24 h. We believe that this additional step allows for more complete hydrolysis and polycondensation reactions and uniform formation of bonds at the molecular level, which will result in the increased homogeneity of the material.

The films were deposited via spin-coating of the final solutions on $Pt/TiO_2/SiO_2/Si$ substrates at 2000 rpm (30 s duration). Sol was applied to the substrate using a precision Cee® 200X centrifugal coating device (SPS-Europe B.V., Putten, The Netherlands). Sol (3 mL for 1 step) was poured onto the substrate at the same time. After this, wet layers were dried at 350 °C (4 min), and then thermally processed at 550 °C (5 min in air) by RTA (rapid thermal annealing). This was repeated 3 times to get the required thickness ($\approx$300 nm). Such obtained layers were then annealed at 200 °C for 20 min, 400 °C for 20 min, and crystallized at 550 °C, 600 °C, 700 °C for 20 min, respectively. The average thickness of the coating per application step is 95–100 nm. $Bi_{0.9}La_{0.1}FeO_3$ (BLFO) thin films were produced using analogous procedure. Lanthanum nitrate, $La(NO_3)_3 \cdot 5H_2O$, was then used for doping. In this case, all metal salts were dissolved separately, including La-nitrate.

X-ray diffraction measurements were done at the University of Aveiro (Aveiro, Portugal) with a Panalytical Empyrean diffractometer (Malvern Panalytic, Almelo, The Netherlands) having Cu Kα1 cathode (λ = 0.15406 nm) and linear PIXEL detector (divergence slit 1/2°). The intensity of the diffractograms was measured by the continuous counting method (step 0.02°, time 200 s) in the 2θ range of 5–90°. Atomic Force Microscopy (AFM) measurements were conducted using a Bruker Multimode instrument (Bruker Nano Surfaces, Santa Barbara) with a Nanoscope (IV) MMAFM-2 unit. The local piezoelectric activity was evaluated using a standard contact mode with Piezoresponse Force Microscopy (PFM) capability. The PFM method is using the inverse piezoelectric effect, which couples electrical and mechanical responses in a sample. The voltage applied to a piezoelectric sample through a conductive tip produces local changes in its dimensions. To detect the polarization distributions the PFM tip is rastered across the film's surface. Details of the PFM measurement procedure are given in [25].

3. Results and Discussion

XRD data for pure and doped BiFeO$_3$ films (Figure 1 and Table 1) annealed at different temperatures were analyzed. Phase analysis was carried out by comparing the experimental interplane distances d with X-ray patterns of the database of the International Diffraction Data Center ICDD PDF2 [26]. Perovskite phase content was obtained by the calculation of intensity ratios of experimental and analytical peaks using Jana 2006 program [27]. In order to get the best approximation of the intensities of the analytical lines of diffractograms, the grating parameters were refined following the procedure described in [27]. The calculation of crystallite sizes was carried out using the Scherer formula [28]. Rietveld refinement used the experimental XRD profile by accounting for the instrumental contribution with the help of Profex interface (BGMN program) [29]. It was found that BiFeO$_3$ films exhibit completely different behaviour as compared to powders [30]. The formation of crystalline structure with high content of a perovskite phase started already at 550 °C. An increase in the synthesis temperature to 600 °C leads to a concurrent increase in the perovskite phase content. Further annealing at 700 °C does not lead to a decrease in the content of the required phase (unlike powders of the same composition) due to possible Bi volatilization [31].

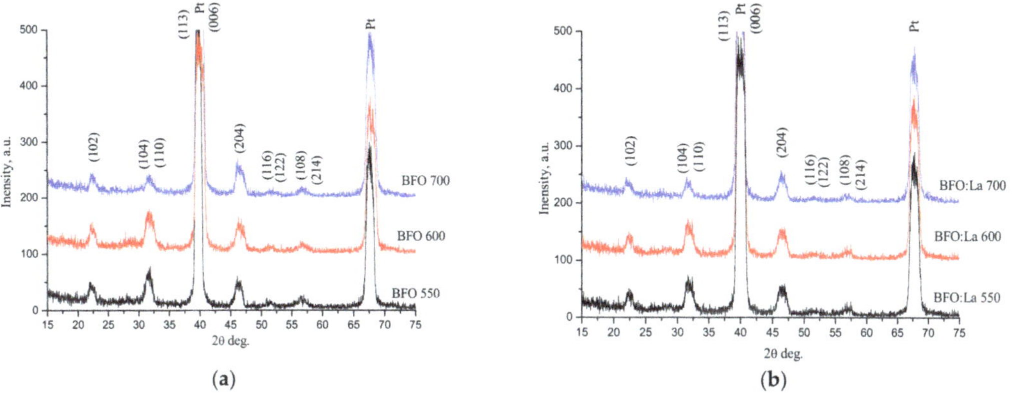

Figure 1. XRD profiles of BiFeO$_3$ (**a**) and Bi$_{0.9}$La$_{0.1}$FeO$_3$ (**b**) sol–gel films.

Table 1. Phase purity and crystallite sizes of BiFeO$_3$ and Bi$_{0.9}$La$_{0.1}$FeO$_3$ thin films annealed under different conditions.

Sample	Crystallization Temperature, °C	Perovskite Phase Content, % (±2%)	Crystallite Size, nm (±0.5 nm)
	550	70	9
BFO	600	94	5
	700	94	5
	550	92	5
BLFO	600	95	6
	700	97	6

As judged from the XRD measurements for BFO and BLFO thin films (Figure 1, Table 1), the content of the perovskite phase in La-doped films is higher than in pure $BiFeO_3$ annealed at the same temperature. This may be explained by the lower degree of volatilization of Bi when it is substituted by La^{3+} [32]. We believe that the addition the lanthanum nitrate leads to a significant decrease in the concentration of intrinsic defects as it was observed in $BiFeO_3$ ceramics (see [32] and references therein). Additionally, this results in the formation of perovskite structure at lower temperatures, e.g., 550 °C, for lanthanum-doped $BiFeO_3$. The observed increase in the content of the perovskite phase was also observed with increasing annealing temperature from 550 °C to 700 °C for both compounds and La-doped films were found to contain significantly less amount of secondary phases (Table 1). This is an expected result taking into account our previous study of BFO and BLFO phase formation [31] and the fact that La addition should promote better chemical homogeneity while simultaneously resulting in the increase of the crystallite size (Table 1).

In addition, it can be noted that the annealing temperature notably affects the orientation of the grains (i.e., the films' texture). For the films annealed at 550 °C and 600 °C, the most intensive peak is at 32 degrees, which corresponds to the (110)/(104) planes of the pseudocubic structure. However, for the films crystallized at 700 °C, the most intensive peak becomes (204). The peak intensity ratios of 550 °C and 600 °C annealed films correspond to a random orientation of the grains [33]. Changing these ratios indicates the predominant orientation of the grains with (102) orientation parallel to the surface of the films. The appearance of (012) texture in the films annealed at high temperature (700 °C) is similar to that observed in ref. [33] and can be related to the growth mechanism, rather than to the effect of the substrate.

The results of the investigation of the surface morphology of the films after annealing at different temperatures are presented in Figure 2 (AFM images) and summarized in Table 2.

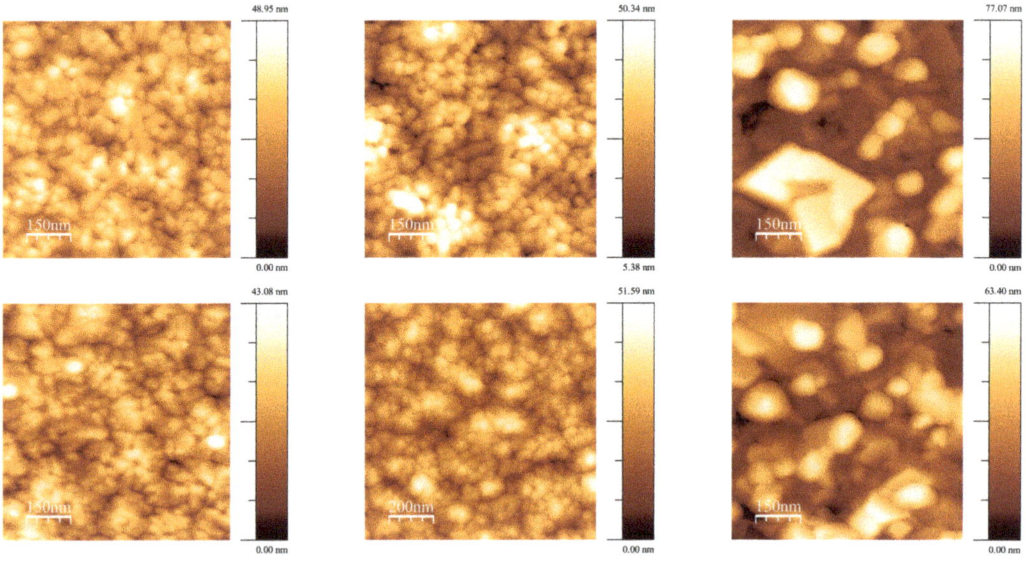

Figure 2. Surface structure of $BiFeO_3$ (**top row**) and $Bi_{0.9}La_{0.1}FeO_3$ (**bottom row**) sol–gel films obtained by AFM at 550 °C (**left column**), 600 °C (**middle column**) and 700 °C (**right column**).

Table 2. Surface parameters of $BiFeO_3$ and $Bi_{0.9}La_{0.1}FeO_3$ sol–gel films annealed at different temperatures.

Sample	Ra, nm	Average Grain Size, nm	Sample	Ra, nm	Average Grain Size, nm
BFO 550 °C	8	90	BLFO 550 °C	4	85
BFO 600 °C	5	80	BLFO 600 °C	4	85
BFO 700 °C	10	100	BLFO 700 °C	9	135

Figure 2 shows the AFM topography images of BFO and BLFO thin films annealed at 550 °C, 600 °C and 700 °C, respectively. The surface morphologies, grain size, geometric shape, homogeneity, and agglomeration of the nanograins can be visualized based on AFM scans. The average grain size was calculated with the Gwyddion program [34]. It can be seen from Figure 2 that, as the annealing temperature increases up to 700 °C, the average grain size increases for both BFO and BLFO thin films. After annealing at 550 °C, the average grain size is about 90 nm, and crystallization at 700 °C results in the average grain size of 100 nm, whereas for some grains it is as large as 300–400 nm (Figure 2). The appearance of relatively large grains for undoped BFO films annealed at 700 °C (Figure 2) is consistent with the observation of increased (102) texture at the same temperatures, so we hypothesize that these grains are (102) oriented. Surface roughness is increased with increasing annealing temperature for both compositions. This is in line with the increased grain size and longer diffusion paths at elevated temperatures. Note that the crystallite size is much less than the grain size and does not depend much on the temperature (except for BFO films annealed at 550 °C with a large amount of secondary phase). It means that the chemical homogeneity within the grains is more or less the same at different annealing temperatures, suggesting that the final crystallization anneal can be done at low temperatures (600 °C).

In order to understand the suitability of deposited BFO-based films for piezoelectric applications PFM method was applied to visualize not only the grain morphology but also their local piezoelectric activity. For grains, in which the polarization is normal to the film's plane, the amplitude of piezoelectric deformation ε measured by PFM is proportional to the longitudinal (or vertical) effective d_{33} coefficient and its phase (close to 0 or to 180°) is a measure of the polarization direction: $\varepsilon_3 = d_{33} \times E_3$ in (*hkl*) grain and $-\varepsilon_3 = d_{33} \times (-E_3)$ in (*-h-k-l*) grain. The 0-degree phase corresponds to the "positive" polar axis and "positive" piezoresponse. If the phase shift is 180°, it indicates the "negative" orientation of the grains and the "negative" piezoresponse. In Figures 3 and 4, the so-called mixed PFM signal is used, in which the phase is reflected in the signal sign. Vertical or out-of-plane (OPP) was measured, which is proportional to the effective piezocoefficient d_{33} [35]. These images are shown for BFO and BLFO thin films annealed at different temperatures. A convenient way to describe the piezosignal distribution is to plot it in the form of histograms showing the strength of the signal as a function of the number of pixels in the entire image [36]. In this way, a complete polarization distribution in as-grown (i.e., not poled by the external electric field) films can be obtained. We note that, due to the small grain size, the contrast is approximately constant within the grain, and no ferroelectric domains are observed. The histograms of the OOP signals taken from the images in Figures 3 and 4 are shown in Figure 5. Comparing the widths of these distributions, one can qualitatively estimate the evolution of piezoelectric activity in the films as a function of annealing temperature (Figures 5 and 6).

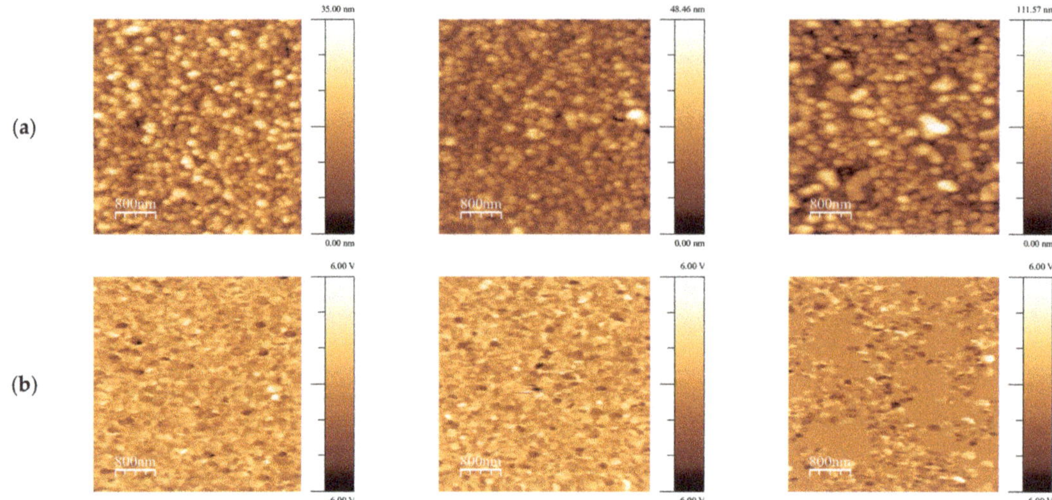

Figure 3. (**a**) Topography, (**b**) out-of-plane (OPP) Piezoresponse Force Microscopy of BiFeO$_3$ sol–gel films annealed at at 550 °C (**left column**), 600 °C (**middle column**) and 700 °C (**right column**).

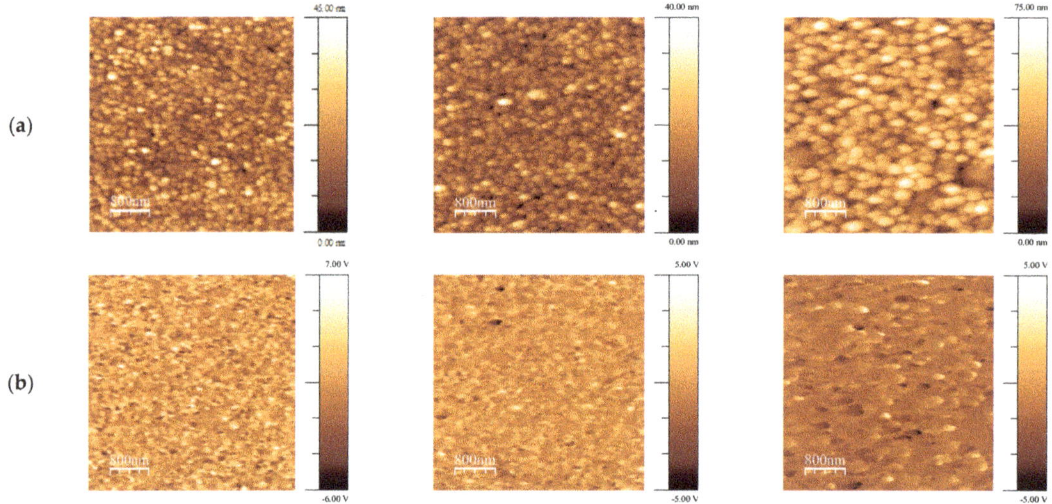

Figure 4. (**a**) Topography, (**b**) OPP PFM of Bi$_{0.9}$La$_{0.1}$FeO$_3$ sol–gel films annealed at 550 °C (**left column**), 600 °C (**middle column**) and 700 °C (**right column**).

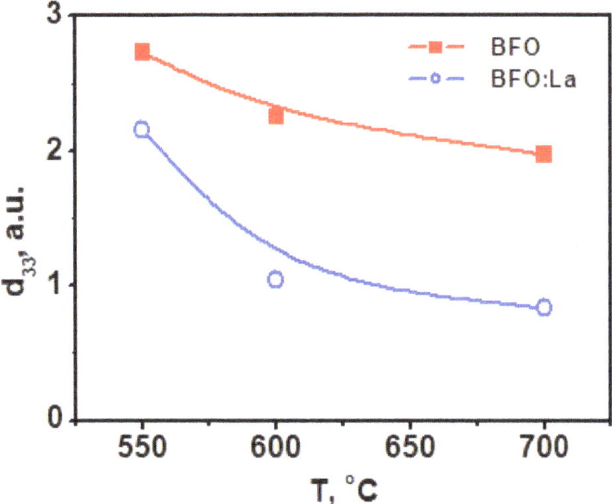

Figure 5. Evolution of the effective piezoresponse of BFO and BFO:La films annealed at different temperatures.

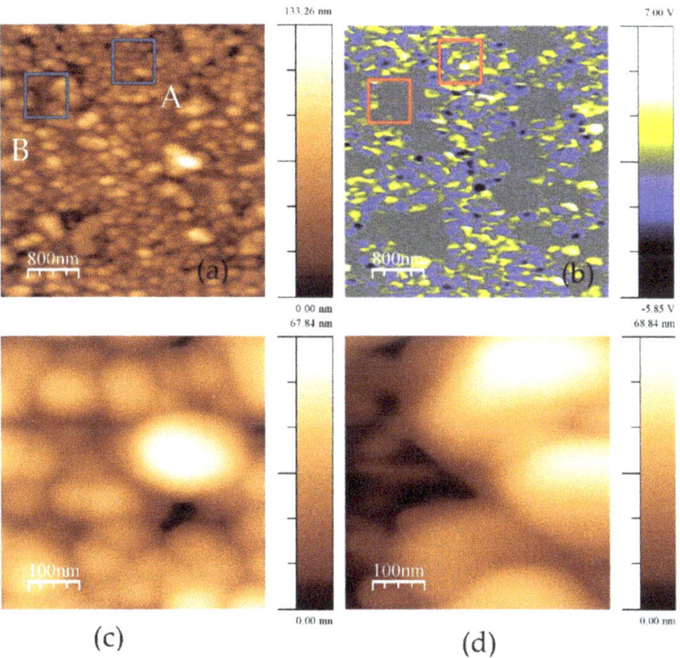

Figure 6. BFO film annealed at 700 °C: (**a**) Topography, (**b**) piezoresponse; and zooms of areas A (**c**,**d**) shown in Figure 7a. Estimated average grain size of A is 100 nm and B—300 nm.

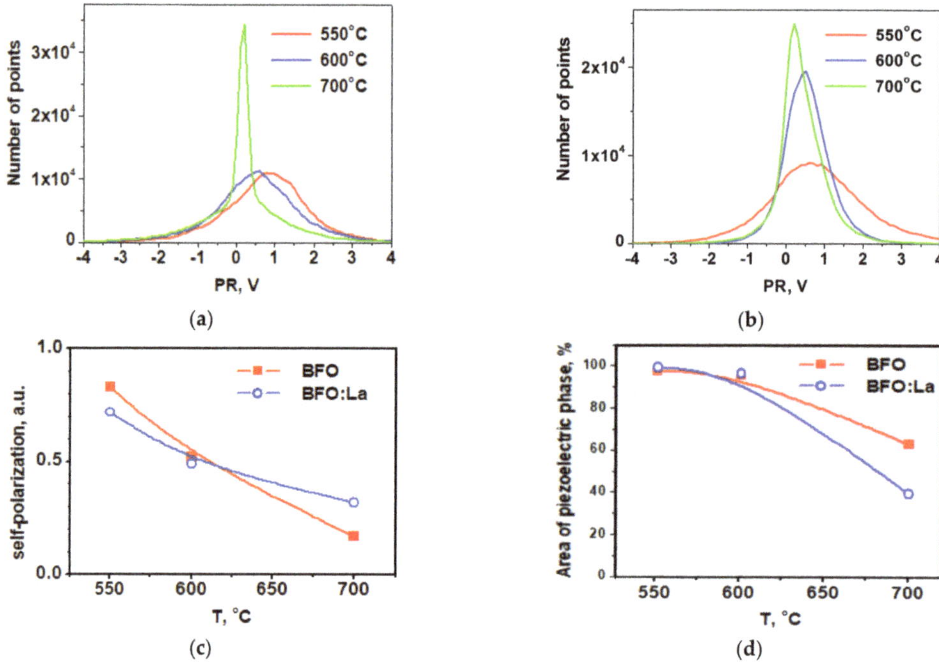

Figure 7. Histograms of the piezoresponse images of BFO (**a**) and BFO:La (**b**) thin films annealed at different temperatures; (**c**) self-polarization in BFO and BFO:La films vs. temperature; (**d**) evolution of areas with piezoresponse after different annealing temperatures.

Figure 5 features a clear decrease in the piezoelectric signal with annealing temperature for both doped and undoped BFO films. In general, doped films are less piezoelectric (Figure 5) because the end member of the solid solution $LaFeO_3$ is centrosymmetric [37].

An interesting observation is that some parts of the films show a zero piezoresponse, i.e., they are not piezoelectric at all. These non-polar or antipolar phases (especially *Pnma* phase) have zero polarization and thus give a zero piezoresponse [38]. Another explanation is the appearance of second phases, for example, $Bi_2Fe_4O_9$, in which polarization should be zero as well [39]. The location of the second phases is most likely on the surface of the films (see [40] and references therein). That is why the relative area of non-piezoelectric inclusions might be higher than the volume concentration of the second phases estimated by XRD. It should be noted that in these areas (where piezoelectric signal is close to zero), the morphology of the grains is different, implying a different growth mechanism. Non-piezoelectric grains are notably larger and the effect is somewhat similar to La addition (Figure 6).

An important property of ferroelectric films is self-polarization, i.e., the ability of the films to exhibit piezoelectric response without any poling [41,42]. In the PFM technique, it is calculated as the relative difference between the "positively" and "negatively" oriented grains/domains on the surface of the films [43]. If the film is strongly self-polarized it can be used as a sensor or actuator without any dubious poling process, which is very difficult to do in the case of microelectromechanical systems (MEMS) [44]. Figure 7a–c shows the evolution of self-polarization with crystallization temperature and La doping. While it decreases with annealing temperature it never goes to zero because T_c in $BiFeO_3$ is higher than the crystallization temperature used in our experiments. It means that the model of polarization by the surface barrier (reported in ref. [40]) can be valid. The effect of La signifies that the Schottky barrier height effectively increases with La doping for the films annealed at higher temperatures. For lower temperatures, this effect is much weaker and might be related to a weaker effect of oxygen vacancies generated by La ([38,45]).

Figure 7d demonstrates that the phase separation is apparently facilitated by La addition. If La concentration in non-piezoelectric grains is greater than 15%, the absence of polar properties can be explained by the appearance of *Pnma* non-polar phase [46]. It can be assumed that during annealing, partial phase separation occurs and the sample is actually a composite comprising polar and non-polar grains similar to that observed in rare-earth-doped BFO ceramics [38]. Figure 7d shows that the large parts of the films do not have a piezoelectric signal after annealing at 700 °C, so 60% of the film surface is not piezoelectric. However, X-ray diffraction measurements based on the comparison of the intensity of the major peaks do not show significant changes (Table 1). Thus PFM is proved to be a very sensitive tool for the monitoring of non-piezoelectric phases on the surface.

4. Conclusions

Pure $BiFeO_3$ and $Bi_{0.9}La_{0.1}FeO_3$ thin films were prepared on $Pt/TiO_2/SiO_2/Si$ substrates by a modified sol–gel technique using a separate hydrolysis procedure. The effects of final crystallization temperature and La doping on phase structure, film surface quality, and nanoscale piezoelectric properties were investigated. Major conclusions of the work can be formulated as follows: (i) La doping and higher crystallization temperature lead to the increase of the grain size and apparent (102) texture of the films. (ii) Simultaneously, a decrease in the average effective piezoelectric coefficient (about 2 times in La-doped films) and an increase in the concentration of surface non-polar phase (up to 60%) are observed. (iii) Phase separation on the surface is related to either the appearance of second phases due to Bi loss at the surface or to the formation of the pseudocubic perovskite phase. (iv) As compared with the undoped $BiFeO_3$, La-doping also increases the average grain size and self-polarization that is important for future piezoelectric applications. Piezoelectric property is shown to be directly connected to the films' growth conditions and doping, emphasizing that both should be thoroughly controlled in order to use BFO-based films in micromechanical applications.

Author Contributions: D.L.K., S.A.K., A.L.K. conceived and designed the experiment; A.V.S., V.V.S. prepared the samples; I.B., V.E.G., S.K. performed the experiments; I.B., S.K., O.P., A.V.S. analyzed the data and prepared the original draft; A.L.K., O.P., D.L.K., S.A.K. reviewed and edited the draft. All authors have read and agreed to the published version of the manuscript.

Funding: This work was developed within the scope of the project CICECO-Aveiro Institute of Materials, refs. UIDB/50011/2020 and UIDP/50011/2020, financed by national funds through the FCT/MCTES. The work was performed within the European Union's Horizon 2020 research and innovation programme under the Marie Sklodowska-Curie grant (agreement No 778070). The project was partly supported by RFBR and BRFBR, project numbers 20-58-0061 and T20R-359, respectively. Part of this work was funded by national funds (OE), through FCT—Fundação para a Ciência e a Tecnologia, I.P., in the scope of the framework contract foreseen in the numbers 4, 5, and 6 of the article 23, of the Decree-Law 57/2016, of August 29, changed by Law 57/2017, of 19 July.

Institutional Review Board Statement: Not applicable.

Informed Consent Statement: Not applicable.

Data Availability Statement: Not applicable.

Conflicts of Interest: The authors declare no conflict of interest.

References

1. Guo, E.J.; Herklotz, A.; Roth, R.; Christl, M.; Das, S.; Widdra, W.; Dörr, K. Tuning the switching time of $BiFeO_3$ capacitors by electrodes' conductivity. *Appl. Phys. Lett.* **2013**, *103*, 022905. [CrossRef]
2. Guo, E.J.; Das, S.; Herklotz, A. Enhancement of switching speed of $BiFeO_3$ capacitors by magnetic fields. *APL Mater.* **2014**, *2*, 096107. [CrossRef]
3. Wu, J.; Fan, Z.; Xiao, D.; Zhu, J.; Wang, J. Multiferroic bismuth ferrite-based materials for multifunctional applications: Ceramic bulks, thin films and nanostructures. *Prog. Mater. Sci.* **2016**, *84*, 335–402. [CrossRef]

4. Chen, Z.; Chen, Z.; Kuo, C.Y.; Tang, Y.; Dedon, L.R.; Li, Q.; Zhang, L.; Klewe, C.; Huang, Y.-L.; Prasad, B.; et al. Complex strain evolution of polar and magnetic order in multiferroic $BiFeO_3$ thin films. *Nat. Commun.* **2018**, *9*, 3764. [CrossRef] [PubMed]

5. Li, J.; Sha, N.; Zhao, Z. Effect of annealing atmosphere on the ferroelectric properties of inkjet printed $BiFeO_3$ thin films. *Appl. Surf. Sci.* **2018**, *454*, 233–238. [CrossRef]

6. Singh, D.; Tabari, T.; Ebadi, M.; Trochowski, M.; Yagci, M.B. Efficient synthesis of $BiFeO_3$ by the microwave-assisted sol-gel method: "A"site influence on the photoelectrochemicalactivity of perovskites. *Appl. Surf. Sci.* **2019**, *471*, 1017–1027. [CrossRef]

7. Machado, P.; Scigaj, M.; Gazquez, J.; Rueda, E.; Sanchez Díaz, A.; Fina, I.; Giber Roca, M.; Puig, T.; Obradors, X.; Campoy-Quiles, M.; et al. Band gap tuning of solution processed ferroelectric perovskite $BiFe_{1-x}Co_xO_3$ thin films. *Chem. Mater.* **2019**, *31*, 947–954. [CrossRef]

8. Karpinsky, D.V.; Troyanchuk, I.O.; Mantytskaya, O.S.; Khomchenko, V.A.; Kholkin, A.L. Structural stability and magnetic properties of $Bi_{1-x}La(Pr)_xFeO_3$ solid solutions. *Solid State Commun.* **2011**, *151*, 1686–1689. [CrossRef]

9. Troyanchuk, I.O.; Karpinsky, D.V.; Bushinsky, M.V.; Khomchenko, V.A.; Kakazei, G.N.; Araujo, J.P.; Tovar, M.; Sikolenko, V.; Efimov, V.; Kholkin, A.L. Isothermal structural transitions, magnetization and large piezoelectric response in $Bi_{1-x}La_xFeO_3$ perovskites. *Phys. Rev. B.* **2011**, *83*, 054109–054115. [CrossRef]

10. Singh Lotey, G.; Verma, N.K. Magnetoelectric coupling in multiferroic Tb-doped $BiFeO_3$ nanoparticles. *Mater. Lett.* **2013**, *111*, 55–58. [CrossRef]

11. Kim, W.-H.; Son, J.H. The effects of La substitution on ferroelectric domain structure and multiferroic properties of epitaxially grown $BiFeO_3$ thin films. *Appl. Phys. Lett.* **2013**, *103*, 132907. [CrossRef]

12. Pan, D.; Zhou, M.; Lu, Z.; Zhang, H.; Liu, J.-M.; Wang, G.-H.; Wan, J.-G. Local magnetoelectric effect in La-doped $BiFeO_3$ multiferroic thin films revealed by magnetic-field-assisted scanning probe microscopy. *Nanoscale Res. Lett.* **2016**, *11*, 318. [CrossRef] [PubMed]

13. Patel, R.; Sawadh, P. Tunable multiferroic properties of cerium doped bismuth ferrite. *Nanosyst. Phys. Chem. Math.* **2019**, *10*, 255–265. [CrossRef]

14. Pradhan, S.K.; Roul, B.K. Effect of Gd doping on structural, electrical and magnetic properties of $BiFeO_3$ electroceramic. *J. Phys. Chem. Solids* **2011**, *72*, 1180–1187. [CrossRef]

15. Sinha, A.K.; Bhushan, B.; Jagannath Sharma, R.K.; Sen, S.; Mandal, B.P.; Meena, S.S.; Bhatt, P.; Prajapat, C.L.; Priyam, A.; Mishra, S.K.; et al. Enhanced dielectric, magnetic and optical properties of Cr-doped $BiFeO_3$ multiferroic nanoparticles synthesized by sol-gel route. *Results Phys.* **2019**, *13*, 102299. [CrossRef]

16. Mukherjee, S.; Gupta, R.; Garg, A.; Bansal, V.; Bhargava, S. Influence of Zr doping on the structure and ferroelectric properties of $BiFeO_3$ thin films. *J. Appl. Phys.* **2010**, *107*, 123535. [CrossRef]

17. Raghavender, A.T.; Hong, N.H. Effects of Mn doping on structural and magnetic properties of multiferroic $BiFeO_3$ nanograins made by sol-gel method. *J. Magn.* **2011**, *16*, 19–22. [CrossRef]

18. Zhang, C.; Song, Y.; Wang, M.; Yin, M.; Zhu, X.; Tian, L.; Wang, H.; Chen, X.; Fan, Z.; Lu, L.; et al. Efficient and flexible thin film amorphous silicon solar cells on nanotextured polymer substrate using sol-gel based nanoimprinting method. *Adv. Funct. Mater.* **2017**, *27*, 1604720. [CrossRef]

19. Dominguez-Trujillo, C.; Peon, E.; Chicardia, E.; Perez, A.; Rodriguez-Ortiz, J.J.; Pavon, J.J.; Garcia-Couce, J.; Galvan, J.C.; Garcia-Moreno, F.; Torresa, Y. Sol-gel deposition of hydroxyapatite coatings on porous titanium for biomedical applications. *Surf. Coat. Technol.* **2018**, *333*, 158–162. [CrossRef]

20. Ibrahim, I.; Dehghanghadikolaei, A.; Advincula, R.; Dean, D.; Luo, A.; Elahinia, M. Ceramic coating for delayed degradation of Mg-1.2Zn-0.5Ca-0.5Mn bone fixation and instrumentation. *Thin Solid Film* **2019**, *687*, 137456. [CrossRef]

21. Sidsky, V.V.; Semchenko, A.V.; Rybakov, A.G.; Kolos, V.V.; Turtsevich, A.S.; Asadchyi, A.N.; Strek, W. La^{3+}-doped $SrBi_2Ta_2O_9$ thin films for FRAM synthesized by sol-gel method. *J. Rare Earths* **2014**, *32*, 277–281. [CrossRef]

22. Perthuis, H.; Colomban, P. Well Densified NASICON Type Ceramics, Elaborated Using Sol-Gel Process and Sintering at Low Temperatures. *Mater. Res. Bull.* **1984**, *19*, 621–631. [CrossRef]

23. Colomban, P. Gel Technology in Ceramics, Glass Ceramics and Ceramic-Ceramic Compositest. *Ceram. Int.* **1989**, *15*, 23–50. [CrossRef]

24. Colomban, P. Chemical Preparation Routes and Lowering the Sintering Temperature of Ceramics. *Ceramics* **2020**, *3*, 312–339. [CrossRef]

25. Balke, N.; Bdikin, I.K.; Kalinin, S.V.; Kholkin, A.L. Electromechanical imaging and spectroscopy of ferroelectric and piezoelectric materials: State-of-the-art and prospects for the future (feature article). *J. Am. Ceram. Soc.* **2009**, *92*, 1629–1647. [CrossRef]

26. International Centre for Diffraction Data. Available online: http://www.icdd.com (accessed on 1 July 2020).

27. Crystallographic Computing System for Standard and Modulated Structures. Available online: http://jana.fzu.cz/workshops/Jana2006%20Cookbook.pdf (accessed on 4 July 2020).

28. Cullity, B.D.; Stock, S.R. *Elements of X-ray Diffraction*, 3rd ed.; Prentice-Hall: Upper Saddle River, NJ, USA, 2001.

29. Open Source XRD Rietveld Refinement. Available online: http://profex.doebelin.org (accessed on 4 July 2020).

30. Lin, Z.; Cai, W.; Jiang, W.; Fu, C.; Li, C.; Song, Y. Effects of annealing temperature on the microstructure, optical, ferroelectric and photovoltaic properties of $BiFeO_3$ thin films prepared by sol-gel method. *Ceram. Int.* **2013**, *39*, 8729–8736. [CrossRef]

31. Semchenko, A.V.; Khakhomov, S.A.; Sidsky, V.V.; Gaishun, V.E.; Kovalenko, D.L.; Strek, W.; Hreniak, D. Structural properties of $BiFeO_3$ and $Bi_{0.9}La_{0.1}FeO_3$ powders synthesized by sol-gel process. In Proceedings of the International Conference on Global Research and Education, Balatonfüred, Hungary, 4–7 September 2019; pp. 113–118.
32. Rojac, T.; Bencan, A.; Malic, M.; Tutuncu, G.; Jones, J.L.; Daniels, J.E.; Damjanovic, D. $BiFeO_3$ ceramics: Processing, electrical, and electromechanical properties. *J. Am. Chem. Soc.* **2014**, *97*, 1993–2011. [CrossRef]
33. Tyholdt, F.; Jørgensen, S.; Fjellvåg, H.; Gunnæs, A.E. Synthesis of oriented $BiFeO_3$ thin films by chemical solution deposition: Phase, texture, and microstructural development. *J. Mater. Res.* **2005**, *20*, 2127–2139. [CrossRef]
34. Gwyddion—Free SPM (AFM, SNOM, NSOM, STM, MFM, . . .) Data Analysis Software. Available online: http://www.gwyddion.net (accessed on 1 September 2020).
35. Bdikin, I.K.; Gracio, J.; Ayouchi, R.; Schwarz, R.; Kholkin, A.L. Local piezoelectric properties of ZnO thin films prepared by RF-plasma-assisted pulsed-laser deposition method. *Nanotechnology* **2010**, *21*, 235703. [CrossRef]
36. Bhargav, K.K.; Ram, S.; Majumdera, S.B. Physics of the multi-functionality of lanthanum ferrite ceramics. *J. Appl. Phys.* **2014**, *115*, 204109. [CrossRef]
37. Karpinsky, D.V.; Troyanchuk, I.O.; Tovar, M.; Sikolenko, V.; Efimov, V.; Kholkin, A.L. Evolution of crystal structure and ferroic properties of La-doped $BiFeO_3$ ceramics near the rhombohedral-orthorhombic phase boundary. *J. Alloys Compd.* **2013**, *555*, 101–107. [CrossRef]
38. Alikin, D.O.; Turygin, A.P.; Walker, J.; Bencan, A.; Malic, B.; Rojac, T.; Shur, V.Y.; Kholkin, A.L. The effect of phase assemblages, grain boundaries and domain structure on the local switching behavior of rare-earth modified bismuth ferrite ceramics. *Acta Mater.* **2017**, *125*, 265–273. [CrossRef]
39. Huang, S.; Qiu, Y.; Yuan, S.L. Enhanced magnetization and electric polarization in $Bi_2Fe_4O_9$ ceramics by magnetic field pre-sintering. *Mater. Lett.* **2015**, *160*, 323–326. [CrossRef]
40. Zhang, Q.; Sando, D.; Nagarajan, V. Chemical route derived bismuth ferrite thin films and nanomaterials. *J. Mater. Chem. C* **2016**, *4*, 4092–4124. [CrossRef]
41. Kholkin, A.L.; Brooks, K.G.; Taylor, D.V.; Hiboux, S.; Setter, N. Self-polarization effect in Pb(Zr,Ti)O_3 thin films. *Integr. Ferroelectr.* **1998**, *22*, 525–533. [CrossRef]
42. Melo, M.; Araújo, E.B.; Shvartsman, V.V.; Shur, V.; Kholkin, A. Thickness effect on the structure, grain size, and local piezoresponse of self-polarized lead lanthanum zirconate titanate thin films. *J. Appl. Phys.* **2016**, *120*, 054101. [CrossRef]
43. Lima, E.C.; Araujo, E.B.; Bdikin, I.K.; Kholkin, A.L. The self-polarization effect in Pb(Zr$_{0.5}$Ti$_{0.5}$)O_3 films with no preferred polarization. *Mater. Res. Bull.* **2012**, *47*, 3548. [CrossRef]
44. Baek, S.H.; Park, J.; Kim, D.M.; Aksyuk, V.A.; Das, R.R.; Bu, S.D.; Felker, D.A.; Lettieri, J.; Vaithyanathan, V.; Bharadwaja, S.S.N.; et al. Giant Piezoelectricity on Si for hyperactive MEMS. *Science* **2011**, *334*, 958–961. [CrossRef] [PubMed]
45. Clark, S.J.; Robertson, J. Band gap and Schottky barrier heights of multiferroic $BiFeO_3$. *Appl. Phys. Lett.* **2007**, *90*, 132903. [CrossRef]
46. Arnold, D.C. Composition-driven structural phase transitions in rare-earth-doped $BiFeO_3$ ceramics: A review. *IEEE Trans. Ultrason Ferroelectr. Freq. Control* **2015**, *62*, 62–82. [CrossRef] [PubMed]

 *materials*

Article

Origin of Mangetotransport Properties in APCVD Deposited Tin Oxide Thin Films

Krunoslav Juraić [1,*], Davor Gracin [1], Matija Čulo [2,3], Željko Rapljenović [2], Jasper Rikkert Plaisier [4], Aden Hodzic [5], Zdravko Siketić [1], Luka Pavić [1] and Mario Bohač [1]

[1]   Ruđer Bošković Institute, Bijenička cesta 54, 10000 Zagreb, Croatia; davor.gracin@irb.hr (D.G.); zdravko.siketic@irb.hr (Z.S.); lpavic@irb.hr (L.P.); mario.bohac@irb.hr (M.B.)

[2]   Institute of Physics, Bijenička cesta 46, 10000 Zagreb, Croatia; mculo@ifs.hr (M.Č.); zrapljenovic@ifs.hr (Ž.R.)

[3]   High Field Magnet Laboratory (HFML-EMFL), Institute for Molecules and Materials, Radboud University, Toernooiveld7, 6525 ED Nijmegen, The Netherlands

[4]   Elettra–Sincrotrone Trieste S.C.p.A., SS 14, km 163.5, 34149 Basovizza, Italy; jasper.plaisier@elettra.eu

[5]   Central European Research Infrastructure Consortium, Strada Statale 14, km 163.5, 34149 Basovizza, Italy; aden.hodzic@ceric-eric.eu

*   Correspondence: kjuraic@irb.hr; Tel.: +385-1-456-0970

Received: 7 October 2020; Accepted: 13 November 2020; Published: 17 November 2020

Abstract: Transparent conducting oxides (TCO) with high electrical conductivity and at the same time high transparency in the visible spectrum are an important class of materials widely used in many devices requiring a transparent contact such as light-emitting diodes, solar cells and display screens. Since the improvement of electrical conductivity usually leads to degradation of optical transparency, a fine-tuning sample preparation process and a better understanding of the correlation between structural and transport properties is necessary for optimizing the properties of TCO for use in such devices. Here we report a structural and magnetotransport study of tin oxide (SnO_2), a well-known and commonly used TCO, prepared by a simple and relatively cheap Atmospheric Pressure Chemical Vapour Deposition (APCVD) method in the form of thin films deposited on soda-lime glass substrates. The thin films were deposited at two different temperatures (which were previously found to be close to optimum for our setup), 590 °C and 610 °C, and with (doped) or without (undoped) the addition of fluorine dopants. Scanning Electron Microscopy (SEM) and Grazing Incidence X-ray Diffraction (GIXRD) revealed the presence of inhomogeneity in the samples, on a bigger scale in form of grains (80–200 nm), and on a smaller scale in form of crystallites (10–25 nm). Charge carrier density and mobility extracted from DC resistivity and Hall effect measurements were in the ranges $1–3 \times 10^{20}$ cm^{-3} and 10–20 cm^2/Vs, which are typical values for SnO_2 films, and show a negligible temperature dependence from room temperature down to −269 °C. Such behaviour is ascribed to grain boundary scattering, with the interior of the grains degenerately doped (i.e., the Fermi level is situated well above the conduction band minimum) and with negligible electrostatic barriers at the grain boundaries (due to high dopant concentration). The observed difference for factor 2 in mobility among the thin-film SnO_2 samples most likely arises due to the difference in the preferred orientation of crystallites (texture coefficient).

Keywords: tin oxide; thin films; atmospheric pressure chemical vapour deposition transport properties; magnetoresistance; impedance spectroscopy; charge carrier mobility

---

1. Introduction

Transparent conductive oxides (TCO) are binary or ternary compounds containing one or two metallic elements. A very good balance of optical and electrical properties characterizes TCO materials.

Widely used TCO include oxides such as ZnO, SnO_2, In_2O_3 doped with metallic elements: Al-doped ZnO (AZO), Sn-doped In_2O_3 (ITO) and F-doped SnO_2 (FTO) [1–3].

Among others, tin oxide (SnO_2), also known as stannic oxide, is an n-type semiconductor (due to oxygen vacancies) with high optical transparency in visible spectral range (>85%) and a wide energy band gap (3.6 eV). It can be found in nature as a mineral known as cassiterite, and it is the main ore of tin [4]. It has a rutile-like crystal structure. The SnO_2 thin films are chemically inert, scratch resistant, and can withstand high temperatures [5].

There are various chemical and physical methods for preparation of pure and doped SnO_2 thin films: chemical vapour deposition, sol gel, spray pyrolysis, electron beam evaporation, vapour deposition, pulsed laser deposition, molecular beam epitaxy, thermal evaporation, reactive evaporation and magnetron sputtering, reactive magnetron sputtering, ion beam deposition [5].

Atmospheric Pressure Chemical Vapour Deposition (APCVD) is a process that enables deposition of vapour species in the form of thin solid films via suitable chemical reactions at atmospheric pressure. By careful choice of the deposition parameters the film properties can be systematically targeted. APCVD is often used in industry for thin film coatings deposition because of reduced costs due to low material consumption, high deposition rates and running costs (compared to low pressure systems) [6].

SnO_2 thin films (pure and doped) have various applications in devices used in daily life: solar energy conversion, flat panel displays, electro-chromic devices, invisible security circuits, LEDs, transparent electrical conductors and non-colouring electrodes, in smart windows, for energy and illumination control, in anti-dazzling rear view windows, and non-emissive displays, low-emittance coatings for energy efficient windows, anti-frost coatings on car windows and transparent electrode for solar cells. In these applications, the film thickness normally lies in the 100–1000 nm range [5].

For photovoltaic application (as transparent front electrode), it is important that SnO_2 is transparent for UV-VIS light while at the same time having very good conductivity. Electrical conductivity can be improved by increasing charge carrier density (doping with foreign atoms) or charge carrier mobility. The most favourable dopants are antimony (SnO_2:Sb) and fluorine (SnO_2:F). Fluorine-doped tin oxide (SnO_2:F, FTO) exhibits good visible transparency owing to its wide band gap, while retaining a low electrical resistivity due to the high carrier concentration caused by the oxygen vacancies and the fluorine dopant. Highers numbers of charge carriers cause lower film transparency. Therefore, better conductivity should be achieved by optimizing charge carrier mobility.

Several different approaches have been reported with respect to how to improve charge carrier mobility of SnO_2, including post-deposition heat treatment, deposition technique, substrate, doping control [7]. For example, charge carrier mobility can be improved by use of highly oriented substrate, use of tin tetrachloride as Sn precursor, higher methanol content.

In our previous publication [8], we reported in detail on the structural properties, examined by XRD, of undoped and doped SnO_2 thin film samples deposited by APCVD with a short discussion about the influence on average transmittance in VIS part of the spectrum and specific surface resistivity at room temperature. In this work, we investigate in more detail the magnetotransport properties of such APCVD deposited SnO_2 thin films in a wide temperature range and relate them with the structural properties. Transport properties were examined in a wide temperature range by impedance spectroscopy, DC resistivity, Hall effect and magnetoresistance. Structure and composition of SnO_2 thin films were examined by Scanning Electron Microscopy (SEM), Grazing Incidence X-Ray Diffraction (GIXRD) and Time-Of-Flight Elastic Recoil Detection Analysis (TOF-ERDA). The analysis of the obtained experimental data shows that the scattering at grain boundaries is the dominant scattering process in this type of SnO_2 samples and the small variation in the charge carrier mobility between the samples most likely stems from the difference in preferred orientation of crystallites (texture coefficient).

2. Materials and Methods

2.1. Thin Film Deposition by APCVD

SnO_2 thin film samples were prepared by the APCVD on soda-lime glass substrates, in an industrial moving belt reactor [9] with constant-temperature zones. The reactor oven is interrupted with two slots perpendicular to the moving direction of the belt, supplied with a setup with a nozzle that enables the vapour or reactants to flow onto the heated glass surface. The reacting gas mixture was $SnCl_4$, H_2O, methanol and oxygen for undoped films, while for doped films, methanol was omitted and the freon gas was added into the vapour mixture. The vapour of reactants was produced in a "bubbler" by passing a carrier gas through the precursors at room temperature (methanol), or moderately heated to 50 °C ($SnCl_4$) or 40 °C (H_2O). The carrier gas for $SnCl_4$, ethanol and H_2O was nitrogen. The temperature of the glass substrate before and after deposition was 590 °C for samples S-590 and B-590, and 610 °C for samples S-610 and B-610 (see Table 1). The glass substrates were loaded by a belt into the furnace for SnO_2 layer deposition. The single-layer deposition duration of 1.5 min and the post-deposition thermal treatment of some 30 min were adjusted by the belt speed. SnO_2 film is formed by a very fast reaction of tetrachloride with water vapour in which methanol is often added as a moderator. For that reason, in the doped samples where the methanol was omitted, the growth rate was about 30% faster, which is similar to the results reported in Ref. [10].

Table 1. Parameters for SnO_2 thin film samples: sample labels, deposition process type, deposition temperature and layer thickness.

Sample Label	Deposition Process Type	Deposition Temperature (°C)	Thickness (nm)
S-590	one-step	590	390
S-610	one-step	610	300
B-590	two-step	590	920
B-610	two-step	610	710

As a result, two types of films were prepared. The first type (S-590 and S-610) was deposited in a one-step process, and the produced samples were intrinsic. The second type (B-590 and B-610) was prepared by depositing the first layer on the glass substrate in the same way as for the first type, while the second (top) layer, which was formed from the solution without methanol and with the addition of fluorine atoms in the form of freon (Chlorofluorocarbon, CFC), was deposited on the already-formed first layer.

2.2. Structural Characterization

2.2.1. Scanning Electron Microscopy (SEM)

Sample surface morphology was analysed using JEOL, JSM 7000F field emission scanning electron microscope (FE-SEM, Zagreb, Croatia). The image acquisition conditions used were: 5 kV, 10 mm working distance, magnification 25k.

2.2.2. Grazing Incidence X-Ray Diffraction (GIXRD)

As-deposited films were thoroughly studied by GIXRD, using the synchrotron radiation source. GIXRD was carried out at the MCX beamline [11] (Elettra synchrotron, Trieste, Italy) with a wavelength of the incident beam of 1.5498 Å (8 keV). The angle of incidence was set to 2.0° (a value much higher than the critical angle for total external reflection for SnO_2 $\alpha_c = \sqrt{2\delta} \approx 0.37°$, calculated from the SnO_2 index of refraction, real part δ [12]). For the critical angle, the beam penetrates 10–20 nm below the surface and gives information about the surface morphology. For the angle of incidence $\alpha_i = 2.0°$ the beam penetrates much deeper and the GIXRD pattern contains the morphological information for the entire SnO_2 layer.

2.2.3. Time-of-Flight Elastic Recoil Detection Analysis (TOF-ERDA)

Atomic content and depth profiles of the elements in the samples were determined using TOF-ERDA. TOF-ERDA measurements were done by 20 MeV $^{127}I^{6+}$ ions with 20° incidence angle toward the sample surface, and TOF-ERDA spectrometer positioned at the angle of 37.5° toward the beam direction. More details about TOF-ERDA setup used in this work can be found in Ref. [13].

2.3. Transport Characterization

2.3.1. Impedance Spectroscopy

Sheet conductivity of SnO_2 thin film samples was measured by impedance spectroscopy (Novocontrol Alpha-N dielectric analyser, Zagreb, Croatia) in the frequency range from 0.01 Hz to 1 MHz, in three subsequent temperature cycles: (i) cooling from 20 °C to −100 °C, (ii) heating from −100 °C to 220 °C, (iii) cooling from 220 °C to 20 °C. The temperature step for all cycles was 40 °C. Sheet conductivity data were normalized to the film thickness (Table 1) to obtain specific conductivity (S/cm).

2.3.2. Magnetotransport

DC resistivity ρ, magnetoresistance and Hall effect measurements were done in the temperature interval from -270 to 27 °C and in magnetic fields up to 5 T. Thin film samples were cut for the Hall-bar geometry with typical dimensions (10 mm × 2 mm × 500 nm). Two current contacts and three pairs of Hall contacts were made by applying silver paint directly to the surface of the films. Magnetoresistance and Hall effect were measured simultaneously at fixed temperatures and magnetic field, B, going from −5 to 5 Tesla. Magnetic field was oriented perpendicular to the current through the sample and perpendicular to the surface of the films. Magnetoresistance data were symmetrized, $V_{xx} = \frac{V_{xx}(+B)+V_{xx}(-B)}{2}$ (V_{xx} is the measured voltage), in order to eliminate the possible mixing of the Hall component and the Hall signal was antisymmetrized, $V_{yx} = \frac{V_{yx}(+B)-V_{yx}(-B)}{2}$, in order to eliminate the possible mixing of the magnetoresistance component. Hall resistance $R_{yx} = V_{yx}/I$ was linear in magnetic field B throughout the whole temperature range for all samples, and the Hall coefficient was obtained as $R_H = \frac{V_{yx}\,t}{I\,B}$, where I is the current and t the sample thickness. The magnetoresistance was determined as standard as $\frac{\Delta\rho}{\rho_0} = \frac{\rho(B)-\rho(0)}{\rho(0)}$.

3. Results and Discussion

3.1. Structural Properties

3.1.1. Scanning Electron Microscopy

In Figure 1, SEM images of SnO_2 thin film samples are presented. Average grain size was calculated from SEM image line profiles using the so-called "linear intercept technique" as follows [14]:

$$\overline{D} = 1.56\frac{C}{MN} \tag{1}$$

where C is the total length of test line used, N is the number of intercepts and M is the SEM image magnification.

There is a significant difference in surface morphology (grain size and shape) for samples deposited in the one-step process—S-590 and S-610 (Figure 1a,b)—and samples deposited a in the two-step process—B-590 and B-610 (Figure 1c,d). Samples B-590 and B-610 have larger grains with sharp edges (pyramidal shape), while samples S-590 and S-610 have smaller grains with a more rounded shape. The samples deposited in the one-step process were deposited with addition of methanol, just as was done in the experiments reported in Refs. [10,15,16], which resulted in smoother films with smaller grains and a higher nucleation density, as reported in Ref. [15,16]. This effect was attributed to the removal of absorbed HCl, which is produced during deposition in reaction of $SnCl_4$ with H_2O. Removal of absorbed HCl leads to an increase in the number of adsorption sites for $SnCl_4$ and H_2O, and consequently to an increase in the micro-grain density and simultaneous decrease of grain size.

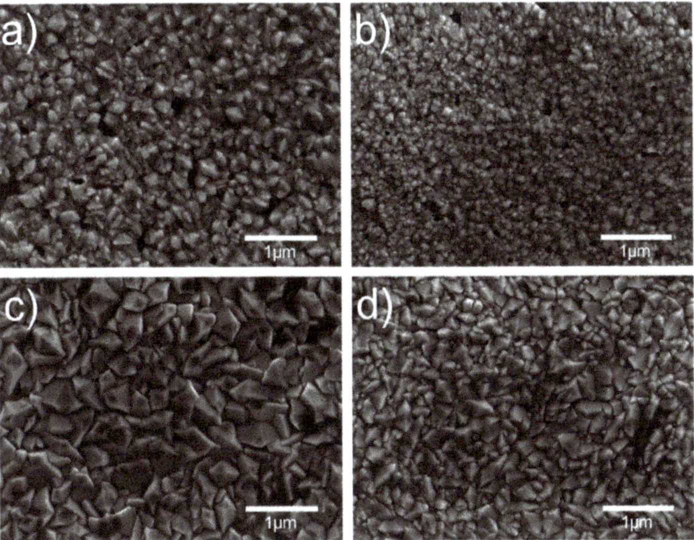

Figure 1. SEM images of SnO_2 samples: (**a**) S-590, (**b**) S-610, (**c**) B-590 and (**d**) B-610. Average grain size calculated using Equation (1) are: 104 nm (S-590), 84 nm (S-610), 190 nm (B-590), 99 nm (B-610).

Higher deposition temperature produces samples with smaller grains (surface roughness) for both single-layer and bi-layer samples. This effect could be a result of competing gas-phase and/or surface reactions, which are a complex function of temperature and composition of reactants [17].

3.1.2. Grazing Incidence X-Ray Diffraction

Figure 2 shows GIXRD patterns of the single-layer and bi-layer SnO_2 thin film samples recorded at the fixed value of incident angle 2.0°. All observed reflections are unambiguously assigned to the tetragonal structure of SnO_2 (space group $P4_2/mnm$, SnO_2 COD database ID: 9009082 [18]), mineralogical name cassiterite). XRD peaks (Figure 2) are fitted to PseudoVoight peak profiles.

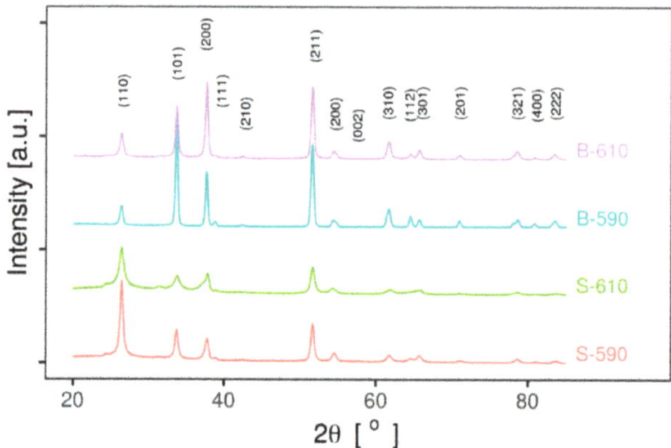

Figure 2. GIXRD diffractograms of SnO_2 thin film samples. All visible peaks are indexed and labelled.

Lattice parameters calculated from peak positions (labelled in Figure 2) are presented. Compared to values from the literature (a = 0.4737(3) nm and c = 0.3186(4) nm), all samples have larger values of the lattice parameter a, and equal values for the lattice parameter c. Variations of lattice parameters are due to the presence of intentional (fluor) and non-intentional (chlorine) dopant atoms with different ionic radii which substitute oxygen atoms in the lattice.

There is also a significant variation in XRD peak intensity ratio from sample to sample, which is also different from the values expected based on the literature. This suggests the presence of preferred orientation (texture) in films. For quantitative analysis the texture coefficient for a selected XRD peak was calculated as intensity ratio of the selected XRD peak and the total sum of all fitted XRD peaks using the following formula [19,20]:

$$TC(h, k, l) = \frac{I(hkl)/I_0(hkl)}{\frac{1}{N}\sum_1^N I(hkl)/I_0(hkl)} \tag{2}$$

where I(h,k,l) is the measured relative intensity of the plane (h k l), I_0(h k l) is the standard intensity of the plane (h k l) taken from literature (COD database, SnO_2 COD ID: 9009082 [18]), and N is the number of reflections included in the calculations. The results for the four most intense XRD peaks are presented in Table 2. It is interesting that samples deposited at the lower temperature (S-590 and B-590) have significantly larger texture coefficients for (110), while B-590 and B-610 samples have larger (200) texture coefficients.

Table 2. Results of GIXRD analysis for SnO_2 samples: lattice parameters (a, c), texture coefficients for four most intense peaks (calculated using Equation (2)), and average crystallite size (calculated using Equation (3)).

Sample	Lattice Parameters		Preferred Orientation Texture Coefficient				Average Crystallite Size (nm)
	a (nm)	c (nm)	(110)	(101)	(200)	(211)	
S-590	0.474(9)	0.318(4)	0.144	0.090	0.104	0.151	17 ± 1
S-610	0.475(7)	0.318(8)	0.143	0.101	0.140	0.136	9 ± 3
B-590	0.475(9)	0.318(7)	0.019	0.141	0.102	0.166	22 ± 3
B-610	0.475(7)	0.318(5)	0.031	0.087	0.173	0.159	20 ± 3

Average crystallite size was estimated using the standard Scherrer Equation [21]:

$$D = \frac{K\lambda}{\beta_{hkl}\cos\theta} \tag{3}$$

where β_{hkl} is XRD line width, D is crystallite size, K is shape factor (0.94) and λ (= 0.154 nm) is the wavelength of Cu K-α radiation. The mean values of average crystallite size obtained by Scherrer equation for the four most intense diffraction peaks are presented in Table 2. The much smaller values for crystallite size compared to grain size estimated from SEM images indicate that grains consist of several crystallites.

Samples deposited in the one-step process have smaller crystallite size calculated from GIXRD and grain size clearly seen from SEM images. Two possible effects/reasons could be responsible for smaller crystallite and grain size in the single-layer samples. First is the smaller thickness of single-layer samples, due to which it can be assumed that there are smaller crystallites and grains at the surface according to the standard thin film growth model [22]. The second one is variation in the deposition temperature. Samples deposited at the lower temperature have larger average crystallite size. Temperature affects a complicated set of chemical reaction mechanisms during the mixing of reactants in vapour phase and at the substrate surface.

3.1.3. TOF-ERDA

TOF-ERDA was employed for the elemental depth profiling of the S-590 and B-590 samples. Due to the overlapping of Sn and scattered I lines in TOF-E spectra (Figure 3a,b), only the first 10^{18} at./cm^2 of sample depth was analysed. Energy spectra belonging to each element were analysed using simulation code Potku [23] (slab analysis) and the Monte Carlo (MC) code CORTEO [24]. Calculated depth profiles (Potku analysis, version 1.1) are presented at Figure 4a,b for S-590 and B-590 respectively. Since the slab analysis does not take into account detector energy resolution and all other contributions to the total energy spread (energy straggling and multiple scattering of incoming and recoiled ions), derived atomic concentrations were used only as input data for the MC simulation. Average atomic concentrations, calculated by MC simulations, are listed in Table 3.

Table 3. Summary of TOF-ERDA elemental analysis (MC simulation). Total atomic content is normalized to 100%.

Sample	H (at.%)	C (at.%)	O (at.%)	F (at.%)	Na (at.%)	Mg (at.%)	Si (at.%)	Cl (at.%)	K (at.%)	Sn (at.%)
S-590	2.8 ± 0.3	1.7 ± 0.2	62 ± 4	-	0.6 ± 0.1	0.18 ± 0.06	1.8 ± 0.2	0.7 ± 0.1	0.25 ± 0.06	29 ± 2
B-590	2.9 ± 0.3	3.2 ± 0.3	62 ± 4	0.9 ± 0.1	0.26 ± 0.06	0.03 ± 0.02	0.11 ± 0.03	0.9 ± 0.1	0.3 ± 0.1	29 ± 2

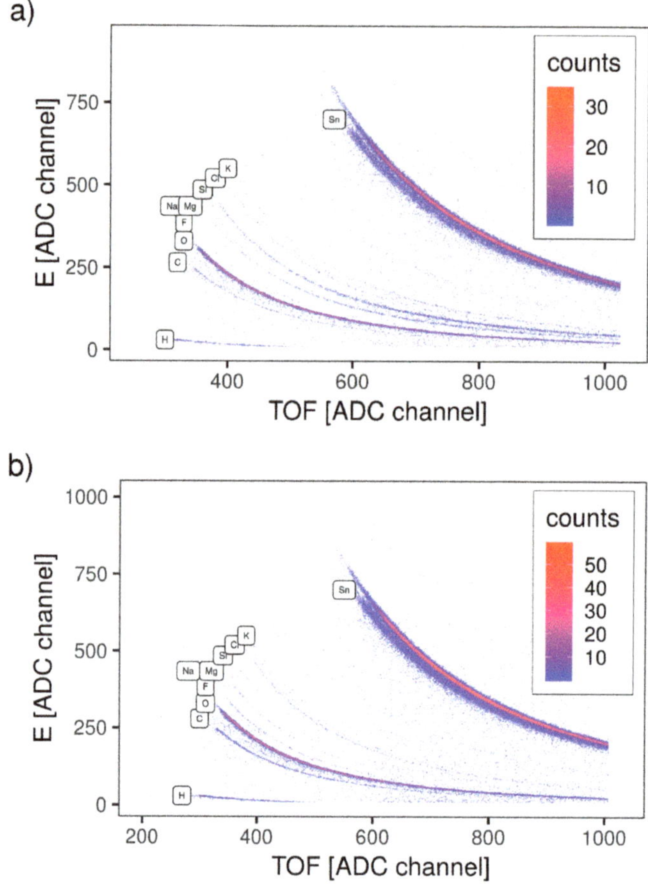

Figure 3. TOF-E map for SnO2 samples: (**a**) S-590 and (**b**) B-590. Traces of all detected elements are labelled.

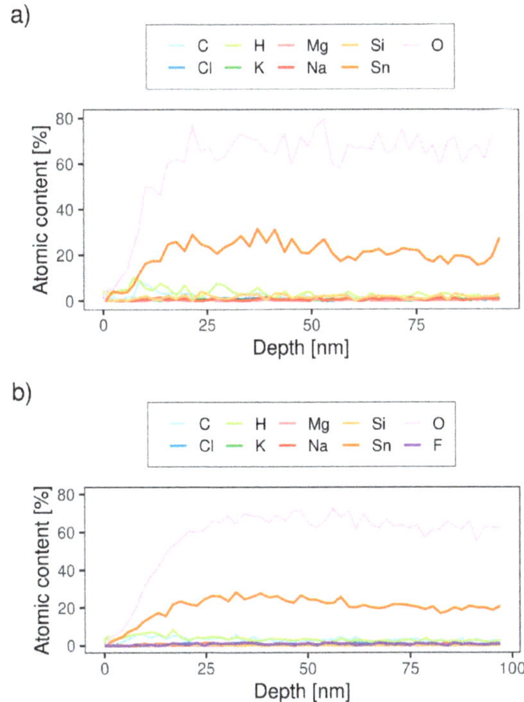

Figure 4. TOF-ERDA elemental depth profile calculated by Potku (slab analysis) for SnO_2 samples: (**a**) S-590 and (**b**) B-590.

The results in Table 3 confirm that the sample deposited in the one-step process (S-590 and S-610) does not contain fluor, as is expected.

For both types of sample, the ratio of Sn and O atoms are stoichiometric (1:2), within the measured error, considering that part of O atoms are bonded to Si in substrate or C atoms at the sample surface. For samples deposited in the two-step process, the concentration of F atoms (dopant) is almost 1%, while for samples deposited in the one-step process, the amount of F atoms is below the detection limit (<0.1 at.%).

Small amounts of Si, K, Mg and Na, visible in TOF-E spectra (Figure 4a,b), could originate from the glass substrate, since the sample area is not fully covered by SnO_2 film. Higher numbers of holes/cracks are expected for thinner samples (single-layer), which could explain the higher contribution of Si, K, Mg and Na in the sample S-590 (Table 3).

3.2. Transport Properties

3.2.1. Impedance Spectroscopy

Impedance spectroscopy results (Figure 5) show that all samples have a very high electrical conductivity, independent of frequency in a wide frequency range indicating fast electronic transport. As expected, samples deposited in the two-step process (B-590 and B-610) have higher conductivity compared to samples deposited in the one-step process at the same temperature because of doping. Only sample S-590 shows a very small dispersion at the highest frequency range. For samples with higher conductivity, a dispersion phenomena is also expected at higher frequencies that are above the limits of the experimental setup. In addition, electrode polarisation effects are not observed for any of the samples, indicating the absence of ion transport contribution to the electrical conductivity.

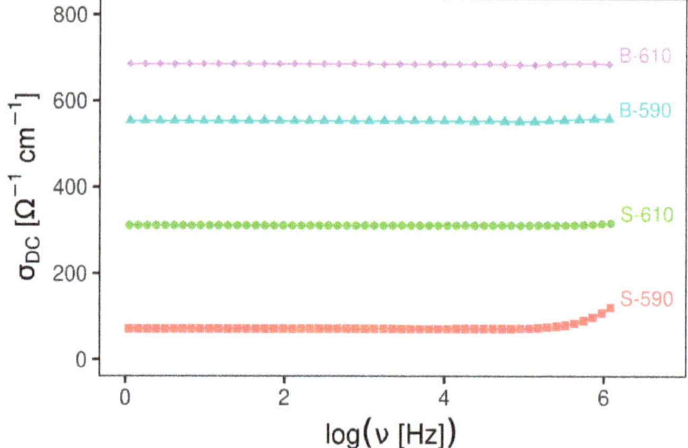

Figure 5. Electrical conductivity for SnO_2 samples as a function of frequency at 20 °C.

3.2.2. Magnetotransport Probe

Figure 6 shows the DC resistivity of the four SnO2 thin films as a function of temperature. Interestingly, for all samples, the DC resistivity has a negligible temperature dependence from room temperature down to −269 °C, indicating a metallic type of charge transport. The high temperature measurements confirmed that the DC resistivity stays almost independent of temperature up to 150 °C. Hall effect measurements showed that the Hall resistivity is linear in a magnetic field up to 5 T and that the carrier density extracted from the Hall coefficient RH is in the range $1–3 \times 10^{20}$ cm^{-3}, which is a typical value for SnO2 films. As expected, RH is negative, indicating n-type free carriers and doped samples (B-590 and B-610) have a higher electron density ($2–3 \times 10^{20}$ cm^{-3}) than the undoped ones (S-590 and S-610) (1×10^{20} cm^{-3}). As can be seen in Figure 7, RH for all samples also shows a negligible temperature dependence, indicating that all samples behave as heavily doped semiconductors and that for both, doped and undoped samples, the Fermi level lies either in the conduction band or in the region where the conduction band is mixed with impurity levels.

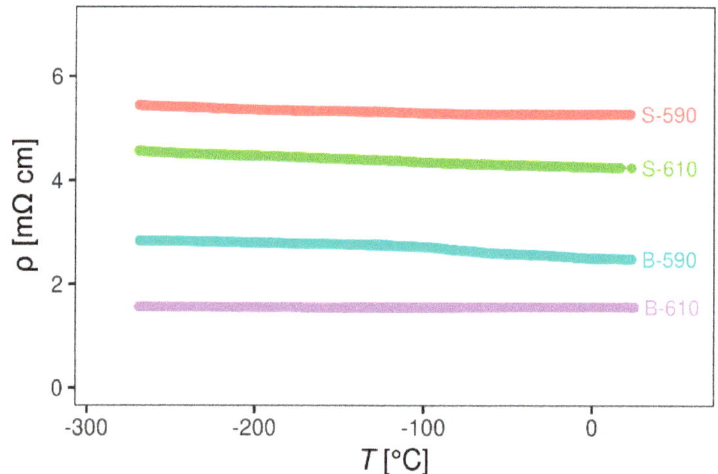

Figure 6. Resistivity of SnO_2 samples as a function of temperature.

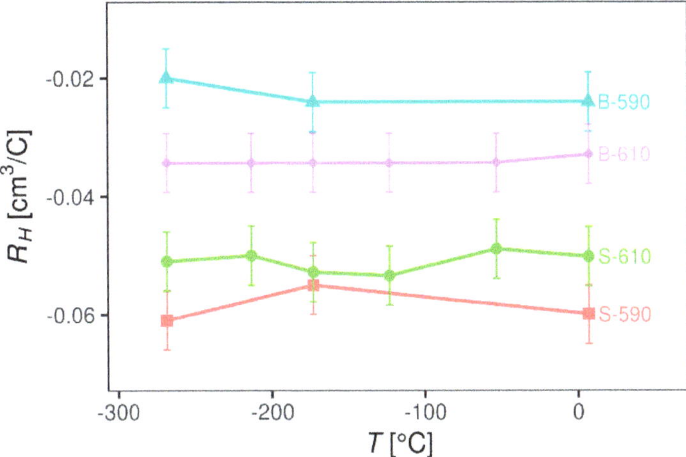

Figure 7. Hall coefficient R_H of SnO_2 samples as a function of temperature.

Having determined the DC resistivity ρ and the Hall coefficient R_H, we are able to calculate the carrier mobility $\mu = R_H/\rho$, which is shown in Figure 8 for all four samples. Remarkably, charge carrier mobility is independent on temperature from room temperature down to −269 °C, in sharp contrast to behaviour found in conventional semiconductors, indicating the dominance of a non-trivial scattering mechanism.

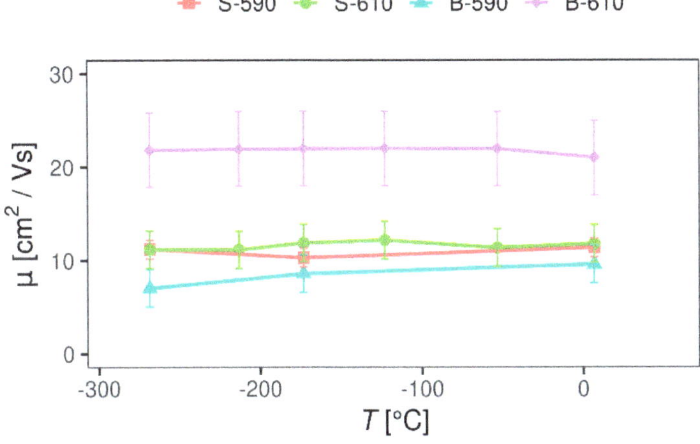

Figure 8. Charge carrier mobility of SnO_2 samples as a function of temperature.

There are several scattering mechanisms influencing the charge carrier mobility in doped semiconductors: electron–phonon scattering, scattering of electrons on ionized impurities, electron–electron scattering, scattering of electrons on neutral impurities, and inter-valley scattering [25]. The last three mechanisms are usually much less pronounced and can generally be ignored for a first approximation. Electron–phonon scattering is a standard scattering mechanism present in all materials and is especially pronounced at high temperatures. Scattering of charge carriers on ionized impurities is usually a second dominant scattering mechanism in doped semiconductors. This is due to the excitation of an electron from the impurity level to the conduction band (n-type) or excitation of an electron from the valence band to the impurity level (p-type) that leaves an uncompensated charge on the impurity. As mentioned earlier, our SnO2 samples are a heavily doped n-type material, intrinsically doped by the oxygen vacancies and extrinsically by the fluorine atoms, so that both electron–phonon and scattering on ionized impurities are expected to play a role in the charge transport. However, both scattering mechanisms show a pronounced temperature dependence [25], in sharp contrast to our negligible temperature dependence of charge carrier mobility (Figure 8), pointing towards the presence of an additional scattering mechanism with basically no temperature dependence.

The charge carrier mobility in polycrystalline samples is known to be determined by grain boundary scattering resulting from the electrostatic charge trapped at the intergrain boundaries, which sets up potential barriers to current flow, although such scattering usually also shows a pronounced temperature dependence. A theoretical model by Prins et al. [26] shows, however, that for certain parameter values the grain–boundary scattering can indeed produce a temperature-independent mobility. The model depends on bulk parameters—the carrier effective mass and the mean free path—and the grain boundary parameters—barrier height, barrier width, and a coefficient of sample inhomogeneity. (The barrier height is defined as the energy difference between the Fermi level and the top of the barrier and the transport in the interior of the grains is separated from the transport across the intergrain boundaries.) The model is tested on five Sb-doped SnO2 thin films with a different doping level. The films with a charge carrier density $>10^{18}$ cm^{-3} showed temperature-independent carrier density (the interior of the grains is degenerately doped) and the sample with the highest dopant concentration showed a negligible temperature dependence of both the carrier density and the carrier mobility over a temperature range of nearly 300 °C, very similar to the behaviour found in our thin films. Moreover, the carrier mobility for the sample with the highest dopant concentration in Ref. [26] was found to be around 18 cm^2/Vs which is very close to the values found in our samples (see Figure 8). Such behaviour is interpreted as originating from the fact that the Fermi level is situated well above the conduction band minimum at the grain boundaries and the negligible electrostatic barriers at the grain boundaries caused by the high dopant concentration (barrier height is negative).

Having established that grain–boundary scattering is responsible for temperature-independent transport in our SnO2 thin films, let us now try to address the small difference in mobility found between the doped and undoped samples prepared at a different deposition temperature. By comparing the grain structure determined by SEM shown in Figure 1 with the charge carrier mobility determined by magnetotransport in Figure 8, no obvious correlation can be found. For example, in contrast to the expectations, the sample with the biggest grains B-590 turns out to have the smallest carrier mobility, while the biggest carrier mobility was found in the sample with the medium grains B-610.

A relatively recent study by Wang et al. [27] indicated the importance of the preferred orientation of crystallites (texture coefficient) in limiting the charge carrier mobility in fluorine-doped thin SnO2 films prepared by APCVD. The main conclusion of this study is that the growth in texture coefficient (110) decreases, while the growth in the texture coefficient (200) increases the carrier mobility in SnO2 films. Looking at Table 2, we can see that the sample B-610, which has the biggest mobility, indeed has the biggest value of the texture coefficient (200), while the sample B-590 with the smallest mobility has the smallest value of the texture coefficient (200), in agreement with the results in Ref. [27]. We can say that the dominant scattering mechanisms in our SnO2 thin films are grain–boundary scattering, responsible for temperature independent carrier mobility, and crystallite scattering, possibly responsible for small differences in carrier mobility among the SnO2 samples prepared under slightly different conditions. More systematic study would be necessary in order to disentangle the contributions coming from the grain–boundary and crystallite–boundary scattering in our SnO2 samples and to establish a direct relationship to the carrier mobility.

Magnetoresistance (Figure 9) of all samples is small (<2%), negative and its value slowly increases with cooling. Negative magnetoresistance is rare in non-magnetic materials and usually has an exotic origin, indicating again that SnO$_2$ samples do not follow simple metallic behaviour. However, negative magnetoresistance is found in impurity conduction of many semiconductors [28–30] and is often attributed to impurity band conduction [31,32], which is in accordance with the conclusions extracted from the temperature dependence of DC resistivity and the Hall coefficient. There are several theoretical models that try to resolve this behaviour; for example, weak localization [31], two band model with sharp mobility edge in the overlap region [32] and spin disorder [33]. The true nature of negative magnetoresistance in our samples is beyond the scope of this paper, but a more comprehensive study of this feature is likely to be part of future publications.

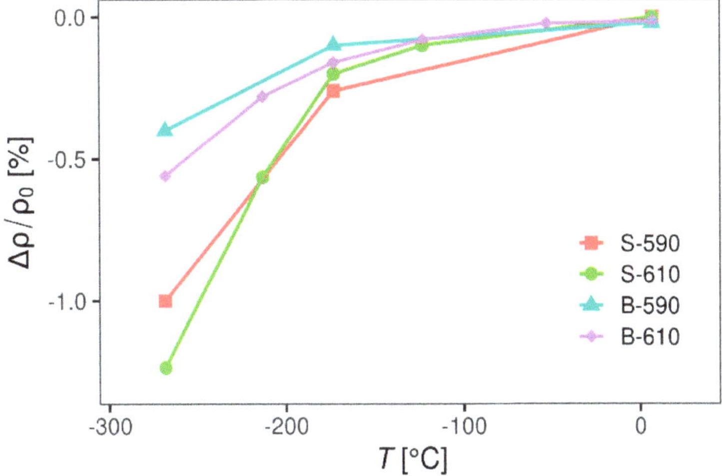

Figure 9. Magnetoresistance of SnO_2 samples as function of temperature. Error bars are omitted in plot because they are order of experimental point symbol size.

4. Conclusions

Undoped and fluorine-doped SnO_2 thin films deposited by APCVD at two different temperatures, 590 °C and 610 °C, were studied by structural and magnetotransport probes. GIXRD revealed the polycrystalline nature of the thin films with the average crystallite size of 10–25 nm, while SEM indicated the presence of an additional inhomogeneity on a bigger scale in form of grains with the average size 80–200 nm. Samples deposited at 590 °C were found to have somewhat bigger grains, while samples deposited at 610 °C showed more (200) preferred orientation of crystallites. Hall effect measurements showed that the carrier density for both undoped and fluorine-doped samples was in the range 1–3×10^{20} cm^{-3} (assuming a substantial level of natural defects and unintentional doping) and has a negligible temperature dependence from room temperature down to −269 °C, indicating that the Fermi level lies either in the conduction band or in the region where the conduction band mixes with the impurity levels. Carrier mobility extracted from the Hall effect and DC resistivity turned out to have values 10–20 cm^2/Vs and to be temperature independent down to −269 °C, showing that the usual scattering of phonons and ionized impurities play a minor role in these samples. Such temperature-independent transport properties are ascribed to the dominant grain boundary scattering, where the interior of the grains is degenerately doped, i.e., the Fermi level is positioned above the conduction band minimum, and due to high dopant concentration, the electrostatic barriers at the grain boundaries are negligible. However, for our samples, there was no obvious correlation between the grain size and the carrier mobility, as the sample with the biggest carrier mobility had medium size grains, while the sample with the smallest carrier mobility had the biggest grains. The highest charge carrier mobility was found in the sample with the largest (200) texture coefficient of crystallites, which is consistent with the results published in Ref. [27]. More systematic studies are needed to separate the influence of grain boundary scattering and scattering on the preferred orientation crystallites and to optimize the transport properties of SnO_2 thin films prepared by APCVD.

Author Contributions: K.J.: sample preparation, SEM, GIXRD experiments and data analysis, writing, original draft preparation, visualization; D.G.: conceptualization, samples preparation; J.R.P. and A.H.: GIXRD experiment; Z.S.: TOF-ERDA experiment and data analysis; M.Č. and Ž.R.: magnetotransport probe experiment, analysis and discussion, draft writing; L.P.: impedance spectroscopy experiment and data analysis; M.B.: SEM experiments and data analysis. All authors have read and agreed to the published version of the manuscript.

Funding: This research was supported by European Regional Development Fund (ERDF) under the (IRI) project "Improvement of solar cells and modules through research and development" (grant number KK.01.2.1.01.0115) and Croatian Science Foundation Project No. IP-2018-01-5246 and No. IP-2013-11-1011 (magnetotransport experiment).

Acknowledgments: We acknowledge Elettra Sincrotrone Trieste for providing access to its synchrotron radiation facilities for GIXRD experiment (MCX beamline). M. Čulo and Ž. Rapljenović thank Đ. Drobac for cutting the samples for Hall bar geometry and B. Radatović for providing a sample holder and helping with high temperature measurements used to verify temperature trend of resistivity above the room temperature.

Conflicts of Interest: The authors declare no conflict of interest.

References

1. Afre, R.A.; Sharma, N.; Sharon, M.; Sharon, M. Transparent conducting oxide films for various applications: A review. *Rev. Adv. Mater. Sci.* **2018**, *53*, 79–89. [CrossRef]

2. Exarhos, G.J.; Zhou, X.-D. Discovery-based design of transparent conducting oxide films. *Thin Solid Films* **2007**, *515*, 7025–7052. [CrossRef]

3. Hosono, H.; Ueda, K. Transparent Conductive Oxides. In *Springer Handbook of Electronic and Photonic Materials*; Kasap, S., Capper, P., Eds.; Springer International Publishing: Cham, Switzerland, 2017; p. 1. ISBN 978-3-319-48931-5.

4. Anthony, J.W. (Ed.) *Handbook of Mineralogy*; Mineral Data Pub: Tucson, AZ, USA, 1990; ISBN 978-0-9622097-0-3.

5. Chen, Z.; Pan, D.; Li, Z.; Jiao, Z.; Wu, M.; Shek, C.-H.; Wu, C.M.L.; Lai, J.K.L. Recent Advances in Tin Dioxide Materials: Some Developments in Thin Films, Nanowires, and Nanorods. *Chem. Rev.* **2014**, *114*, 7442–7486. [CrossRef] [PubMed]

6. Aliano, A.; Cicero, G.; Nili, H.; Green, N.G.; García-Sánchez, P.; Ramos, A.; Lenshof, A.; Laurell, T.; Qi, A.; Chan, P.; et al. Atmospheric Pressure Chemical Vapor Deposition (APCVD). In *Encyclopedia of Nanotechnology*; Bhushan, B., Ed.; Springer Netherlands: Dordrecht, Germany, 2012; p. 146. ISBN 978-90-481-9750-7.

7. Calnan, S.; Tiwari, A.N. High mobility transparent conducting oxides for thin film solar cells. *Thin Solid Film.* **2010**, *518*, 1839–1849. [CrossRef]

8. Djerdj, I.; Gracin, D.; Juraić, K.; Meljanac, D.; Bogdanović-Radović, I.; Pletikapić, G. Structural analysis of monolayered and bilayered SnO_2 thin films. *Surface Coat. Technol.* **2012**, *211*, 24–28. [CrossRef]

9. Gordon, R. Chemical vapor deposition of coatings on glass. *J. Non Cryst. Solids* **1997**, *218*, 81–91. [CrossRef]

10. Volintiru, I.; de Graaf, A.; van Deelen, J.; Poodt, P. The influence of methanol addition during the film growth of SnO_2 by atmospheric pressure chemical vapor deposition. *Thin Solid Film.* **2011**, *519*, 6258–6263. [CrossRef]

11. Rebuffi, L.; Plaisier, J.R.; Abdellatief, M.; Lausi, A.; Scardi, P. MCX: A synchrotron radiation beamline for X-ray diffraction line profile analysis. *Z. Für Anorg. Und Allg. Chem.* **2014**, *640*, 3100–3106. [CrossRef]

12. Henke, B.L.; Gullikson, E.M.; Davis, J.C. X-Ray Interactions: Photoabsorption, scattering, transmission, and reflection at E = 50–30,000 eV, Z = 1–92. *At. Data Nucl. Data Tables* **1993**, *54*, 181–342. [CrossRef]

13. Siketić, Z.; Radović, I.B.; Jakšić, M. Development of a time-of-flight spectrometer at the Ruđer Bošković Institute in Zagreb. *Nucl. Instrum. Methods Phys. Res. Sect. B Beam Interact. Mater. Atoms* **2008**, *266*, 1328–1332. [CrossRef]

14. Wurst, J.C.; Nelson, J.A. Lineal intercept technique for measuring grain size in two-phase polycrystalline ceramics. *J. Am. Ceram. Soc.* **1972**, *55*, 109. [CrossRef]

15. Matsui, Y.; Mitsuhashi, M.; Yamamoto, Y.; Higashi, S. Influence of alcohol on grain growth of tin oxide in chemical vapor deposition. *Thin Solid Film.* **2007**, *515*, 2854–2859. [CrossRef]

16. Matsui, Y.; Mitsuhashi, M.; Goto, Y. Early stage of tin oxide film growth in chemical vapor deposition. *Surf. Coat. Technol.* **2003**, *169–170*, 549–552. [CrossRef]

17. Ghoshtagore, R.N. Mechanism of CVD Thin Film SnO_2 Formation. *J. Electrochem. Soc.* **1978**, *125*, 110. [CrossRef]

18. Gražulis, S.; Daškevič, A.; Merkys, A.; Chateigner, D.; Lutterotti, L.; Quirós, M.; Serebryanaya, N.R.; Moeck, P.; Downs, R.T.; Le Bail, A. Crystallography open database (COD): An open-access collection of crystal structures and platform for world-wide collaboration. *Nucleic Acids Res.* **2012**, *40*, D420–D427. [CrossRef] [PubMed]

19. Shanthi, S.; Anuratha, H.; Subramanian, C.; Ramasamy, P. Effect of fluorine doping on structural, electrical and optical properties of sprayed SnO_2 thin films. *J. Cryst. Growth* **1998**, *194*, 369–373. [CrossRef]

20. Barrett, C. *Structure of Metals*; Horney Press: New York, NY, USA, 2007; ISBN 978-1-4437-3140-9.

21. Patterson, A.L. The scherrer formula for X-Ray particle size determination. *Phys. Rev.* **1939**, *56*, 978–982. [CrossRef]

22. Kaiser, N. Review of the fundamentals of thin-film growth. *Appl. Opt.* **2002**, *41*, 3053–3060. [CrossRef]

23. Arstila, K.; Julin, J.; Laitinen, M.I.; Aalto, J.; Konu, T.; Kärkkäinen, S.; Rahkonen, S.; Raunio, M.; Itkonen, J.; Santanen, J.-P.; et al. Potku—New analysis software for heavy ion elastic recoil detection analysis. *Nucl. Instrum. Methods Phys. Res. Sect. B Beam Interact. Mater. Atoms* **2014**, *331*, 34–41. [CrossRef]

24. Schiettekatte, F. Fast Monte Carlo for ion beam analysis simulations. *Nucl. Instrum. Methods Phys. Res. Sect. B Beam Interact. Mater. Atoms* **2008**, *266*, 1880–1885. [CrossRef]

25. Kasap, S.O.; Capper, P. (Eds.) *Springer Handbook of Electronic and Photonic Materials*; Springer: New York, NY, USA, 2006; ISBN 978-0-387-26059-4.

26. Prins, M.W.J.; Grosse-Holz, K.-O.; Cillessen, J.F.M.; Feiner, L.F. Grain-boundary-limited transport in semiconducting SnO_2 thin films: Model and experiments. *J. Appl. Phys.* **1998**, *83*, 888–893. [CrossRef]

27. Wang, J.T.; Shi, X.L.; Liu, W.W.; Zhong, X.H.; Wang, J.N.; Pyrah, L.; Sanderson, K.D.; Ramsey, P.M.; Hirata, M.; Tsuri, K. Influence of preferred orientation on the electrical conductivity of fluorine-doped tin oxide films. *Sci. Rep.* **2015**, *4*, 3679. [CrossRef] [PubMed]

28. Fritzsche, H.; Lark-Horovitz, K. Electrical properties of p-type indium antimonide at low temperatures. *Phys. Rev.* **1955**, *99*, 400–405. [CrossRef]

29. Woods, J.F.; Chen, C.Y. Negative magnetoresistance in impurity conduction. *Phys. Rev.* **1964**, *135*, A1462–A1466. [CrossRef]

30. Sasaki, W. Negative magnetoresistance in the metallic impurity conduction of n-type germanium. *J. Phys. Soc. Jpn.* **1965**, *20*, 825–833. [CrossRef]

31. Dauzhenka, T.A.; Ksenevich, V.K.; Bashmakov, I.A.; Galibert, J. Origin of negative magnetoresistance in polycrystalline SnO_2 films. *Phys. Rev. B* **2011**, *83*, 165309. [CrossRef]

32. Giovannini, B.; Hedgcock, F.T. Influence of temperature on the two band model for negative magnetoresistance in heavily doped semiconductors. *Solid State Commun.* **1972**, *11*, 367–370. [CrossRef]

33. Boon, M.R. Negative Magnetoresistance in Doped Semiconductors. *Phys. Rev. B* **1973**, *7*, 761–762. [CrossRef]

Article

Optimizing the Properties of InGaZnO$_x$ Thin Film Transistors by Adjusting the Adsorbed Degree of Cs$^+$ Ions

He Zhang [1,2], Yaogong Wang [1,2,*], Ruozheng Wang [1,2], Xiaoning Zhang [1,2] and Chunliang Liu [1,2]

[1]   Key Laboratory of Physical Electronics and Devices, Xi'an Jiaotong University, Ministry of Education, Xi'an 710049, China
[2]   School of Electronic and Information Engineering, Xi'an 710049, China
*   Correspondence: wyg008@xjtu.edu.cn

Received: 27 May 2019; Accepted: 16 July 2019; Published: 18 July 2019

Abstract: To improve the performance of amorphous InGaZnO$_x$ (a-IGZO) thin film transistors (TFTs), in this thesis, Cs$^+$ ions adsorbed IGZO (Cs-IGZO) films were prepared through a solution immersion method at low temperature. Under the modification of surface structure and oxygen vacancies concentrations of a-IGZO film, with the effective introduction of Cs$^+$ ions into the surface of a-IGZO films, the transfer properties and stability of a-IGZO TFTs are greatly improved. Different parameters of Cs$^+$ ion concentrations were investigated in our work. When the Cs$^+$ ions concentration reached 2% mol/L, the optimized performance Cs-IGZO TFT was obtained, showing the carrier mobility of 18.7 cm^2 V^{-1} s^{-1}, the OFF current of 0.8×10^{-10} A, and the threshold voltage of 0.2 V, accompanied by the threshold voltage shifts of 1.3 V under positive bias stress for 5000 s.

Keywords: low-temperature fabrication; ions adsorption; IGZO TFTs; device performance

1. Introduction

In the last decade, amorphous InGaZnO$_x$ (a-IGZO) thin film transistors (TFTs) have been extensively researched due to their outstanding performance, including superior carrier mobility (μ_{FE}), low subthreshold swing ($S.S$), large switching current radio (I_{ON}/I_{OFF}), and high transparency under visible light [1–3]. It showed great significance in the applications of next-generation electronic devices such as displays [4,5], sensors [6,7], and memories [8,9], especially in wearable and flexible devices like foldable displays and e-paper [10–12]. Since the wearable and flexible devices should be prepared onto a polymeric substrate which cannot afford the high temperature, the a-IGZO TFTs need to be prepared at a low temperature for use in wearable and flexible applications.

In recent years, solution process and magnetron sputtering have been commonly used to achieve low temperature prepared a-IGZO TFTs. However, a-IGZO prepared by the solution process shows disadvantages such as lots of film defects, large roughness, and poor uniformity of large areas—inducing low carrier mobility of a-IGZO TFTs (generally < 10 cm^2 V^{-1} s^{-1}), which cannot satisfy the requirements of high-definition flexible display applications [13–16]. The studies also showed that most of the sputtered a-IGZO TFTs always need an annealing process to guarantee their high performance, since unannealed a-IGZO fabricated by magnetron sputtering still have many defects inside [17–19]. Although some scholars have fabricated a-IGZO TFTs by magnetron sputtering without annealing, the low mobility (<10 cm^2 V^{-1} s^{-1}) and stability (threshold voltage shifts > 5 V) of TFTs determined that they cannot afford the requirement of flexible devices [20–22]. Until now, it is still an open question to obtain a high-performance a-IGZO TFT at low temperature without an annealing process.

Herein, a kind of Cs$^+$ ion adsorption was used to modify the surface of a-IGZO film to improve the performance of a-IGZO TFTs. The excellent electrical properties of a-IGZO TFT were obtained under low-temperature conditions, without post-annealing, by adjusting the adsorption degree of Cs^+ ions. Compared with traditional a-IGZO TFT fabricated by a high-temperature annealing process, excellent properties were observed in the optimized Cs^+ ions adsorbed IGZO (Cs-IGZO) TFT. The high carrier mobility of 18.7 cm^2 V^{-1} s^{-1}, the low threshold voltage (V_{th}) of 0.2 V, and the low V_{th} shifts of 1.3 V under positive bias stress for 5000 s reveal the outstanding performance of the Cs-IGZO TFT.

2. Materials and Methods

The Bottom-gate staggered Cs-IGZO TFTs were fabricated and shown in Figure 1. In Cs-IGZO TFTs fabrication, heavily doped p-type silicon (P++-Si) was chosen to be the substrate and gate electrode. A layer of SiN$_x$, with a thickness of 50 nm, was deposited on the Si wafer by plasma enhanced chemical vapor deposition (PECVD), serving as dielectric layer. Then, a-IGZO film with a thickness of 50 nm was deposited onto SiN$_x$ by magnetron sputtering. The working pressure during a-IGZO film deposition was maintained at 5 mTorr with a mixture of Ar (24.3 sccm) 97% and O$_2$ (0.7 sccm) 3% gases. The DC power and sputtering time were 80 W and 30 min, respectively. After IGZO deposition, the wet etching technique was applied to form active islands of TFTs.

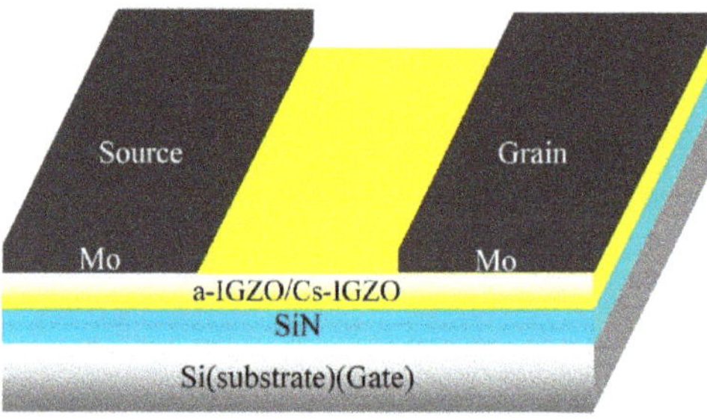

Figure 1. Structure of Cs$^+$ ions adsorbed IGZO (Cs-IGZO) thin film transistor (TFT) and IGZO TFT.

The Cs-IGZO film was formed by immersing the prepared substrate into CsHCO$_3$ solution. To investigate the influence of the Cs$^+$ ions concentration in the performance of Cs-IGZO film and TFTs, four different concentrations of CsHCO$_3$ were prepared as shown in Table 1. The substrate was immerged into CsHCO$_3$ solution for 60 min under 75 °C to ensure the sufficient adsorption of Cs$^+$ ions into IGZO film. Then, Cs-IGZO films were cleaned by deionized water and dried in nitrogen. Finally, a 50 nm thick layer of molybdenum (Mo) was deposited through a shadow mask to form source (S) and drain (D) electrodes, which also defined the channel length as 500 μm.

Table 1. Name of Cs-IGZO films.

Name of Cs-IGZO Film	A1	A2	A3	A4
Concentration of CsHCO$_3$ Solution (% mol/L)	0.5	1	2	3

Transfer properties and stability under positive gate bias of Cs-IGZO TFTs were measured by a semiconductor parameter analyzer (Keithley 4200SCS, Cleveland, OH, USA) in a dark box at room temperature. Hall mobility (μ_{Hall}), carrier concentration, and resistivity of Cs-IGZO films were obtained by Hall Effect. The surface morphologies and the chemical composition of Cs-IGZO films were obtained by Atomic Force Microscope (AFM, INNOVA, Billerica, MA, USA), X-Ray Diffraction (XRD, D8 ADVANCE A25, Bruker, Karlsruhe, Germany), and X-ray Photoelectron Spectroscopy (XPS, Thermo Fisher ESCALAB Xi+, Waltham, MA, USA).

3. Results and Discussion

3.1. Electrical Characteristics of Cs-IGZO TFTs

The transfer properties of Cs-IGZO TFTs from sample A1 to A4 were measured and plotted in Figure 2. In the measurement of Cs-IGZO TFTs transfer properties, the drain–source voltage (V_{DS}) is 10 V, with sweeping the source–gate voltage (V_{GS}) from -5 to $+20$ V. V_{th}, μ_{FE}, and $S.S$ of TFTs are extracted by the transfer properties of the TFTs and represented in Table 2, respectively. As observed with Figure 2 and Table 2, the V_{th} of samples A1 to A4 gradually reduce from 7.9 V to -1.7 V, which indicates that the switching performance of Cs-IGZO TFTs can be adjusted with the changing of Cs^+ ions concentration. The μ_{FE} of samples A1 to A4 increase from 8.6 cm^2 V^{-1} s^{-1} to 21.5 cm^2 V^{-1} s^{-1}, indicating that the transport speed of electrons of a-IGZO film can be significantly improved by increasing the Cs^+ ions concentration. The $S.S$ of samples A1 to A4 decreases from 0.28 V/decade to 0.22 V/decade, which means the defects inside the TFTs can be tapered by increasing the Cs^+ ions concentration. In addition, the OFF current (I_{OFF}) of samples A1 to A4 are calculated to be 5.6 $\times$ 10^{-11} A, 3.5 $\times$ 10^{-10} A, 0.8 $\times$ 10^{-10} A, and 6.7 $\times$ 10^{-8} A, which indicates that the drive characteristics of Cs-IGZO TFTs can be deteriorated by increasing the concentration of Cs^+ ions.

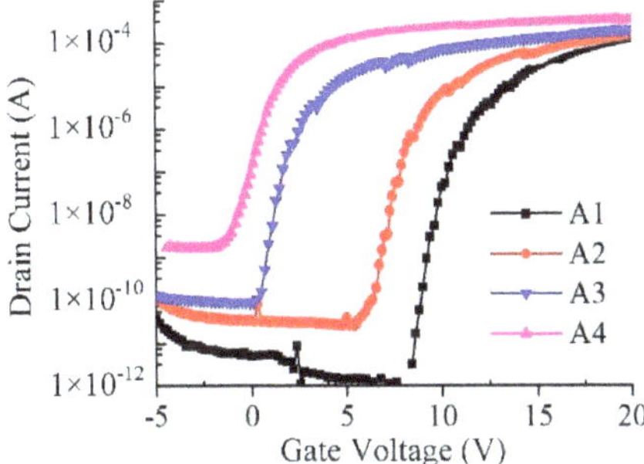

Figure 2. Transfer performance of sample A1 to A4.

Table 2. V_{th} and $S.S$ value of samples A1 to A4.

Parameter/Sample	V_{th} (V)	μ_{FE} (cm^2 V^{-1} s^{-1})	$S.S$ (V/decade)
A1	7.9	8.6	0.28
A2	5.7	13.1	0.25
A3	0.2	18.7	0.23
A4	-1.7	21.5	0.22

In general, the oxygen molecules in air ambient can adsorb onto a-IGZO when a-IGZO is under positive bias and exposed in the atmosphere. The formation of oxygen adsorption is shown in the following equation:

$$O_2\,(gas) + e^- \leftrightarrow 2O^-\,(solid) \tag{1}$$

This adsorption of oxygen can change the physicochemical properties of the a-IGZO film, causing the V_{th} positive shifts (ΔV_{th}) of the a-IGZO TFT under working conditions, thereby reducing the stability of the IGZO TFT. Therefore, in order to investigate the effect of Cs^+ ions adsorbed degree on the stability of Cs-IGZO TFTs, the ΔV_{th} of samples A1 to A4 under positive bias ($V_{GS} = 10$ V) for different time are calculated and represented in Figure 3, respectively. As observed with Figure 3, sample A1 has the largest ΔV_{th} (3.9 V), represented by the black dash line. The ΔV_{th} of samples A1 to A3 are gradually reduced, wherein the ΔV_{th} of samples A2 (2.8 V) and A3 (1.3 V) are represented by red and blue lines, respectively. These results indicate the stability of Cs-IGZO TFTs can be optimized by increasing the Cs^+ ions adsorbed degree of IGZO film. In addition, sample A4 has a similar ΔV_{th}

to sample A3 of 1.2 V, represented by the purple line, indicating that the stability of Cs-IGZO TFTs are no longer optimized by increasing the Cs^+ ions adsorbed degree of IGZO film when the $CsHCO_3$ concentration reached 2% mol/L.

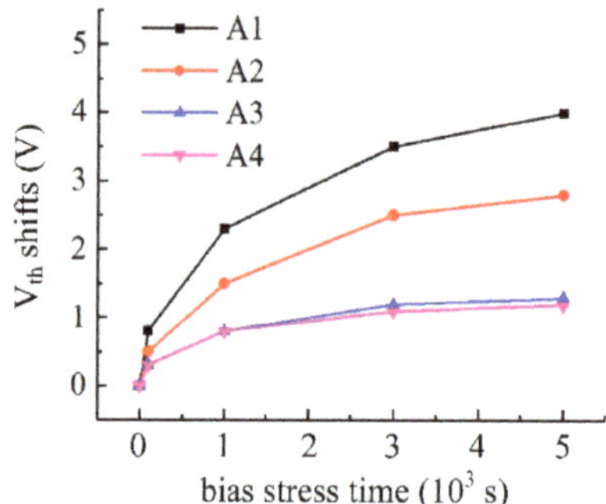

Figure 3. V_{th} shift under positive bias stress of samples A1 to A4.

3.2. Surface Structure of Cs-IGZO Films

The surface crystal structures of samples A1 to A4 were measured by XRD, the results are shown in Figure 4.

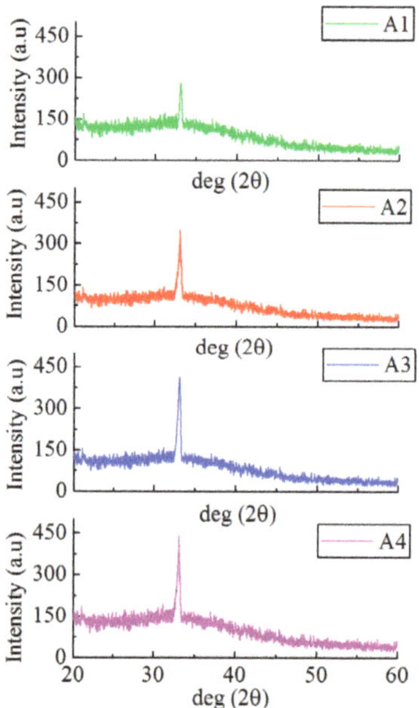

Figure 4. XRD results of samples of A1, A2, A3, and A4.

As observed with Figure 4, the surfaces of the samples A1 to A4 all have a diffraction peak at a diffraction angle of 33.1°, indicating that the samples A1 to A4 all exhibit a single crystal state. The intensity of the diffraction peaks in samples A1 to A4 gradually increased, indicating that the surface crystallization strength of the Cs-IGZO film can be promoted by increasing the degree of adsorption of Cs^+ ions, thereby obtaining the more regular surface structure of Cs-IGZO films. Since the diffraction peak at a diffraction angle of 33.1° of Cs-IGZO films corresponds to (101) orientation of indium, the XRD results of samples A1 to A4 also indicate that the regular-arranged In atoms are the primary cause of surface crystallization of Cs-IGZO films [23].

The surface morphology of samples A1 to A4 are measured using AFM and shown in Figure 5a–d, respectively. As observed with Figure 5, the surface morphology of samples A1 to A4 gradually changed from random arrangement to linear arrangement, which not only indicates the Cs^+ ions adsorption could obtain a-IGZO films with regular surface structure, but also indicates that the surface structure of IGZO thin films tend to be more regular by increasing the Cs^+ ion adsorbed degree.

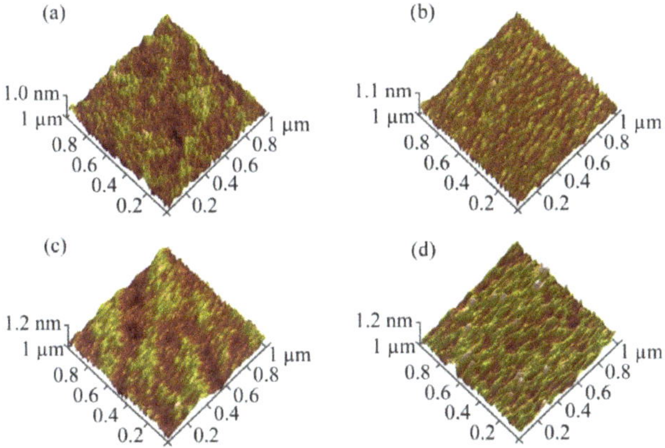

Figure 5. AFM images of samples of (**a**) A1, (**b**) A2, (**c**) A3, and (**d**) A4.

According to the XRD and AFM results, the movement states of carriers in Cs-IGZO films with different Cs^+ ions adsorption are given and represented in Figure 6. Figure 6a represents the electrons transport path inside the Cs-IGZO film when the Cs^+ ions are barely adsorbed, like sample A1. In this state, most of the metal–oxide molecules in the surface of the Cs-IGZO films are randomly arranged. Since the electrons move in the oxygen vacancies in Cs-IGZO films, the electrons need to move around each molecule, therefore the transmission paths represent the irregular way.

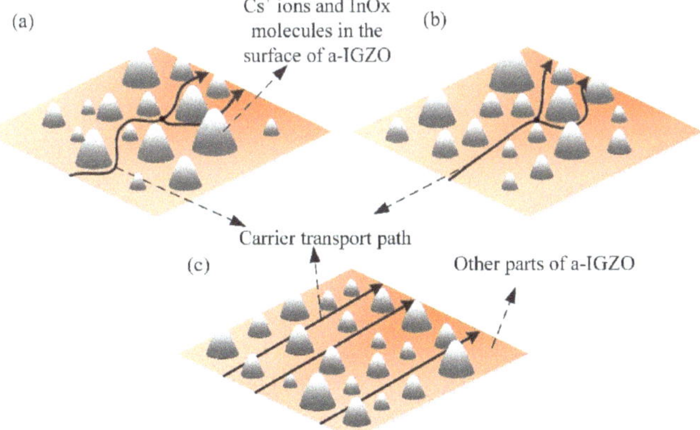

Figure 6. Diagram of carrier transport path in IGZO film. (**a**) barely Cs^+ ions adsorbed IGZO film. (**b**) part of Cs^+ ions adsorbed IGZO film. (**c**) large amount of Cs^+ ions adsorbed IGZO film.

When the Cs^+ ions adsorbed degree is gradually increased, some of the InO_x molecules on the surface of the Cs-IGZO film begin to have the regular arrangement, like samples A2. The schematic of the carrier transport path in this kind of situation is represented by Figure 6b. As adsorbed from Figure 6b, the electrons can be transmitted linearly around the regularly distributed InO_x molecules but are still transmitted in an irregular way around the randomly distributed other metal–oxide molecules, so their electron transport paths consist of half straight lines and half irregular lines.

As shown in Figure 6c, with the further increase of Cs^+ ions adsorbed degree, most of the InO_x molecules in the surface of the Cs-IGZO film are completely regular-arranged, like sample A3 and A4. At this time, the electrons do not need to move to bypass any molecules, so their transmission paths show a straight line. In this situation, the electrons' transmission path will be greatly shortened, thereby improving the electrons' transport speed efficiency, which can effectively increase the carrier mobility of the IGZO film.

3.3. Chemical Studies on Cs-IGZO Films

In order to determine the adsorbed degree of Cs^+ ions in Cs-IGZO films, the Cs_{3d} spectrum of high-resolution XPS spectra for samples A1 to A4 are measured and represented in Figure 7a. The Cs_{3d} spectra of samples A1 to A4 are indicated by the green line, blue line, orange line, and purple line, respectively. As observed in Figure 7a, Cs_{3d} spectrum of all the four samples exhibit two distinct peaks at binding energies 724.8 eV and 738.6 eV, which represent the chemical state of $Cs_{3d}^{2/3}$ and $Cs_{3d}^{5/2}$, respectively, ensuring the presence of Cs^+ ions. At the same time, their peak intensities increase with the increasing of $CsHCO_3$ concentration, indicating that the Cs^+ ion content in samples A1 to A4 is gradually increased. Combined with XPS analysis software, the content of Cs^+ ions of Cs-IGZO films can be calculated. The Cs^+ content in samples A1 to A4 is 0.49%, 0.78%, 1.21%, and 1.46%, indicating the content of Cs^+ ions in Cs-IGZO films increases with the $CsHCO_3$ concentration rising.

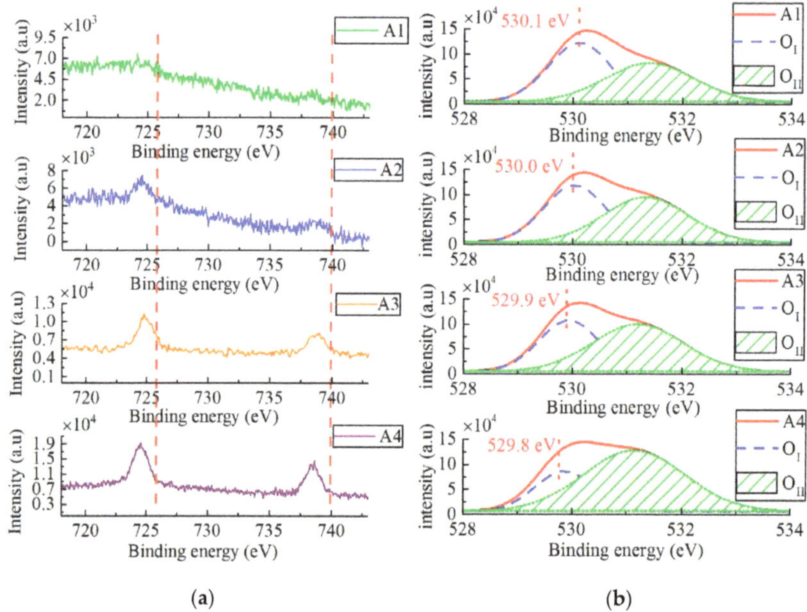

Figure 7. (a) High resolution XPS spectrum of Cs3d for sample A1 to A4. (b) High resolution XPS spectra of O1s for sample A1 to A4.

The O 1s spectra of samples A1 to A4 are analyzed by XPS to investigate the effect of Cs^+ ions concentration on the characteristics of Cs-IGZO films. Generally, the O 1s spectrum of IGZO film consists of lattice oxygen (O_I), which represents the oxygen in oxide lattices; and vacancy oxygen (O_{II}), which exists in oxygen deficient. The O_I spectrum locates at lower binding energy with a peak of around 529.6 eV; and the O_{II} spectrum locates at higher binding energy with a peak of around 531.3 eV. According to the positions of O_I and O_{II} spectra, the O_{1s} spectra of samples A1 to A4 are divided into two spectra, respectively. Figure 7b shows the O_{1s} spectra and fitting curves of O_I and O_{II} of samples A1 to A4.

As observed in Figure 7b, the peak intensities in O_{II} spectra of samples A1 to A4 gradually increase, indicating that the oxygen vacancy concentrations of IGZO film can be significantly improved by increasing the Cs^+ ions adsorbed degree. The area percentage under the peak of O_{II} spectrum in O_{1s} spectra ($O_{II}/(O_I+O_{II})$) of samples A1 to A4 are calculated to be 48.1%, 51.9%, 52.7%, and 72.3%, respectively, ascertaining that the high adsorption degree of Cs^+ ions can generate more oxygen vacancies inside the a-IGZO film. Since the oxygen vacancy work as the donor impurities of IGZO films, this increased oxygen vacancy of Cs-IGZO films is attributed to the improvement of electrical properties of TFTs, including the negative shift of V_{th} and the increase of μ_{FE} and the carrier concentration.

According to the physical and chemical analysis above, the reasons for the performance improvement of Cs-IGZO TFTs are discussed. The negative shift of V_{th} of samples A1 to A4 are attributed to the low work function of cesium and the increased oxygen vacancies of IGZO films. When Cs^+ ions gradually adsorb in IGZO film, the low work function of cesium (2.17 eV) can gradually eliminate the Schottky barrier which exists in the interface between the source/drain electrodes and the active layer, resulting in the negative shift of V_{th} of samples A1 to A4. When concentration of oxygen vacancy increased, the conduction band inside IGZO films gradually moved down, reducing the distance between the Fermi level and the conduction band, resulting in the negative shift of V_{th}.

The highly improved μ_{FE} is attributed to the increased oxygen vacancy and the regular-arranged surface structure of IGZO films. Since each oxygen vacancy generated within IGZO can provide two free electrons for the conduction band, the increased oxygen vacancies inside IGZO are usually accompanied by an improvement in the carrier concentration and carrier mobility of IGZO TFTs. The regular-arranged surface structure of IGZO films lead the carriers and can be transmitted in a straight line in the surface of IGZO film, effectively improving the carrier mobility of IGZO TFTs.

The increase of I_{OFF} is mainly attributed to the surface single crystallization of the Cs-IGZO film by adsorbing the Cs^+ ions. The surface single crystallization can produce the grain boundaries in the surface of Cs-IGZO. Under this circumstance, parts of electrons will flow along the grain boundary when electrons flow on the surface of Cs-IGZO, causing the growth of I_{OFF} in Cs-IGZO TFT. As the degree of crystallization of Cs-IGZO surface increases, more electrons can flow between the grain boundaries, resulting in the improvement of I_{OFF} in Cs-IGZO TFTs.

The reason for the optimized stability of Cs-IGZO TFTs is analyzed and shown in Figure 8.

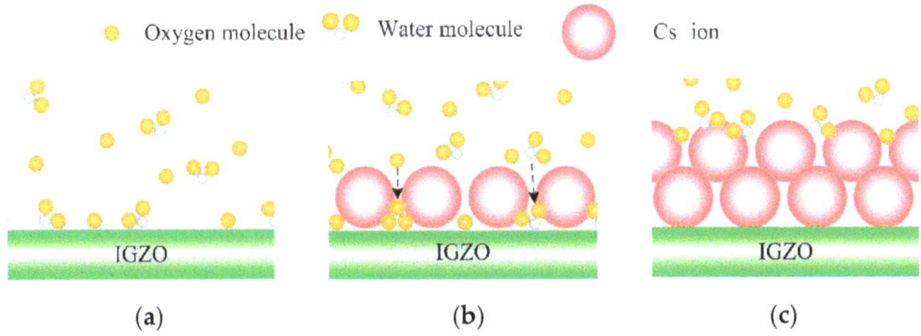

Figure 8. Schematic diagram of adsorption of H_2O and oxygen molecules in the surface of Cs-IGZO and amorphous $InGaZnO_x$ (a-IGZO) films. (**a**) barely Cs^+ ions adsorbed IGZO film. (**b**) part of Cs^+ ions adsorbed IGZO film. (**c**) large amount of Cs^+ ions adsorbed IGZO film.

Figure 8a is the schematic which shows the adsorbed state of oxygen molecules and water molecules in air on the bare Cs^+ ions adsorbed IGZO film. In this situation, oxygen molecules and water molecules can be adsorbed in the surface of the IGZO film without any barrier, resulting in the positive shift of V_{th} of IGZO TFTs. When Cs^+ ions are gradually adsorbed into IGZO, the Cs^+ ions can block the contact of IGZO with molecules in the air, greatly reducing the adsorption of oxygen molecules and water molecules on the surface of IGZO, thereby effectively improving the stability of IGZO TFT, as shown in Figure 8b. However, since Cs^+ ions have a large ionic radius, oxygen molecules and water molecules can pass through the interspace between the Cs^+ ions to adsorb in the IGZO film, therefore small amount of Cs^+ ion adsorption cannot completely isolate the contact between IGZO and molecules in air. As observed in Figure 8c, the oxygen molecules and water molecules cannot pass through the Cs^+ ions barrier to adsorb in the surface of IGZO film when a large number of Cs^+ ions are adsorbed in IGZO film. In this state, the IGZO TFT has superior stability, and its stability cannot be improved as the Cs^+ ions concentration increases. Combined with the electrical properties of samples A1 to A4, it can be concluded that sample A3 is the optimized Cs-IGZO TFT which has the optimized performance.

3.4. Comparison between Cs-IGZO TFT and Annealed IGZO TFT

The performance of the optimized Cs-IGZO TFT (sample A3) and a-IGZO TFT which was annealed at 400 °C for 30 min in N_2 atmosphere is compared. The transfer properties of sample A3 and annealed a-IGZO TFT are represented in Figure 9, and the V_{th}, μ_{FE}, and $S.S$ are calculated and represented in Table 3. As observed in Figure 9 and Table 3, sample A3 has a similar V_{th} to that of annealed a-IGZO TFT which is 0.2 V, and its μ_{FE} and $S.S$ are superior to that of a-IGZO TFT, indicating the transfer properties of optimized Cs-IGZO TFT are superior to that of annealed IGZO TFT.

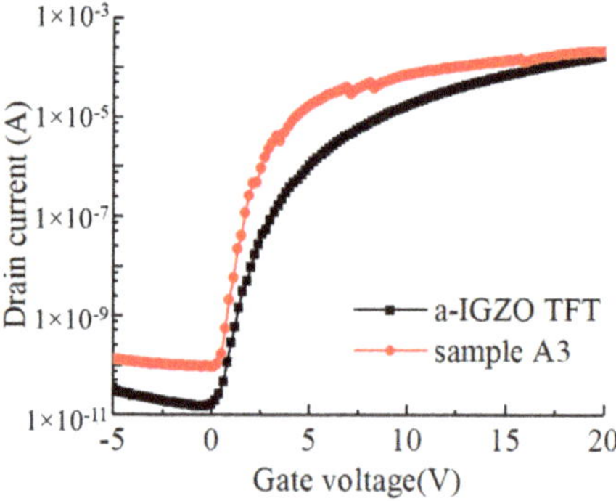

Figure 9. Comparison of electrical properties between sample A3 and annealed IGZO TFT.

Table 3. Electrical properties of optimized Cs-IGZO TFT and a-IGZO TFT.

Samples/Parameter	V_{th} (V)	μ_{FE} (cm^2 V^{-1} s^{-1})	$S.S$ (V/decade)
Cs-IGZO TFT	0.2	18.7	0.23
a-IGZO TFT	0.2	12.6	0.24

The V_{th} shifts under gate bias of sample A3 and annealed a-IGZO TFT are calculated and shown in Figure 10.

As observed in Figure 10, it is obvious that the ΔV_{th} of sample A3 (1.3 V) is smaller than that of annealed a-IGZO TFT (1.4 V). These results indicate that the Cs-IGZO TFT fabricated at a low temperature has the potential to replace the annealed a-IGZO TFTs, and is suitable for use in flexible substrates.

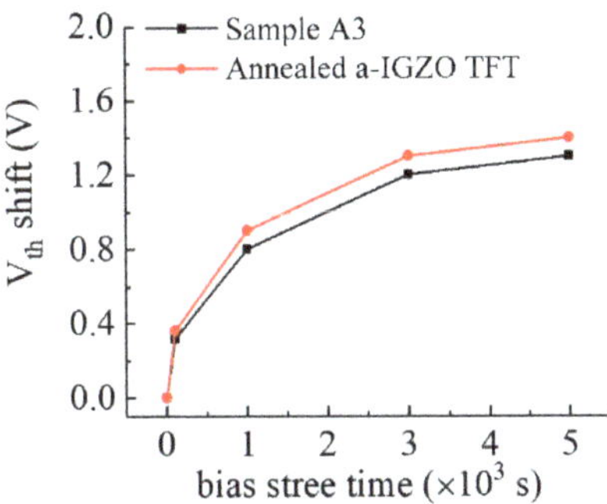

Figure 10. V_{th} shifts of optimized Cs-IGZO TFT and a-IGZO TFT.

4. Conclusions

In summary, Cs^+ ions were adsorbed into IGZO film by $CsHCO_3$ solution immersion method at low temperature, and had the benefit of improving IGZO TFTs performance. Four kinds of $CsHCO_3$ concentrations were selected to optimize the transfer property and stability of Cs-IGZO TFTs. According to the electrical measurements of Cs-IGZO TFTs, it was found that the Cs-IGZO TFT with $CsHCO_3$ concentration of 2% in water solution had the optimized electrical properties, including high μ_{FE} of 18.7 cm^2 V^{-1} s^{-1}, small threshold voltage shifts of 1.3 V, and low OFF current of 0.8×10^{-10} A. These superior performances of optimized Cs-IGZO TFT were attributed to the change of surface structure and oxygen vacancy concentrations of IGZO film, by appropriate Cs^+ ions adsorbed into IGZO film. Compared with the traditional a-IGZO TFTs fabricated through annealing process, the optimized Cs-IGZO TFT had superior mobility and comparable device stability, which might be applicable to future flexible electronics.

Author Contributions: Conceptualization, H.Z., C.L. and X.Z.; methodology, H.Z.; software, Y.W.; validation, H.Z., R.W. and Y.W.; formal analysis, H.Z.; investigation, H.Z.; resources, X.Z.; data curation, H.Z.; writing—original draft preparation, H.Z.; writing—review and editing, Y.W. and X.Z.; visualization, R.W.; supervision, R.W.; project administration, X.Z.; funding acquisition, X.Z. and Y.W.

Funding: The authors would like to acknowledge the financial support provided by National Natural Science Foundation of China (61771382, 51807156), China Postdoctoral Science Foundation (2017M623174), and Projects of International Cooperation and Exchanges Shaanxi Province (2018KW-034).

Conflicts of Interest: The authors declare no conflict of interest.

References

1. Yabuta, H.; Sano, M.; Abe, K.; Aiba, T.; Den, T.; Kumomi, H.; Nomura, K.; Kamiya, T.; Hosono, H. High-mobility thin-film transistor with amorphous ingazno4 channel fabricated by room temperature rf-magnetron sputtering. *Appl. Phys. Lett.* **2006**, *89*, 112123. [CrossRef]
2. Kamiya, T.; Nomura, K.; Hosono, H. Present status of amorphous In-Ga-Zn-O thin-film transistors. *Sci. Technol. Adv. Mater.* **2010**, *11*, 044305. [CrossRef] [PubMed]
3. Liu, H.C.; Lai, Y.C.; Lai, C.C.; Wu, B.S.; Zan, H.W.; Yu, P.; Chueh, Y.L.; Tsai, C.C. Highly effective field-effect mobility amorphous ingazno tft mediated by directional silver nanowire arrays. *ACS Appl. Mater. Interfaces* **2015**, *7*, 232–240. [CrossRef] [PubMed]
4. Liu, F.M.; Wu, Y.C.; Hsu, Y.J.; Yu, M.J.; Im, J.S.; Lu, P.Y. P-169: A 31-inch 4k2k top-emission oled display using good uniformity and long-term reliability top-gate self-aligned IGZO TFTs. *SID Symp. Dig. Tech. Pap.* **2018**, *49*, 1796–1799. [CrossRef]
5. Mo, Y.G.; Kim, M.; Kang, C.K.; Jeong, J.H.; Park, Y.S.; Choi, C.G.; Kim, H.D.; Kim, S.S. Amorphous-oxide TFT backplane for large-sized AMOLED TVs. *J. Soc. Inf. Disp.* **2011**, *19*, 16–20. [CrossRef]

6. Liang, L.; Zhang, S.; Wu, W.; Zhu, L.; Xiao, H.; Liu, Y.; Zhang, H.; Javaid, K.; Cao, H. Extended-gate-type igzo electric-double-layer tft immunosensor with high sensitivity and low operation voltage. *Appl. Phys. Lett.* **2016**, *109*, 173501. [CrossRef]

7. Zan, H.W.; Li, C.H.; Yeh, C.C.; Dai, M.Z.; Meng, H.F.; Tsai, C.C. Room-temperature-operated sensitive hybrid gas sensor based on amorphous indium gallium zinc oxide thin-film transistors. *Appl. Phys. Lett.* **2011**, *98*, 253503. [CrossRef]

8. Hung, M.F.; Wu, Y.C.; Chang, J.J.; Chang Liao, K.S. Twin thin-film transistor nonvolatile memory with an indium–gallium–zinc–oxide floating gate. *IEEE Electron Device Lett.* **2013**, *34*, 75–77. [CrossRef]

9. Lee, G.G.; Tokumitsu, E.; Yoon, S.M.; Fujisaki, Y.; Yoon, J.W.; Ishiwara, H. The flexible non-volatile memory devices using oxide semiconductors and ferroelectric polymer poly(vinylidene fluoride-trifluoroethylene). *Appl. Phys. Lett.* **2011**, *99*, 12901. [CrossRef]

10. Petti, L.; Münzenrieder, N.; Vogt, C.; Faber, H.; Büthe, L.; Cantarella, G.; Bottacchi, F.; Anthopoulos, T.D.; Tröster, G. Metal oxide semiconductor thin-film transistors for flexible electronics. *Appl. Phys. Rev.* **2016**, *3*, 021303. [CrossRef]

11. Lee, S.M.; Shin, D.; Yun, I. Degradation mechanisms of amorphous inGaZnOthin-film transistors used in foldable displays by dynamic mechanical stress. *IEEE Trans. Electron Devices* **2017**, *64*, 170–175. [CrossRef]

12. Ning, H.; Zeng, Y.; Kuang, Y.; Zheng, Z.; Zhou, P.; Yao, R.; Zhang, H.; Bao, W.; Chen, G.; Fang, Z.; et al. Room-temperature fabrication of high-performance amorphous In-Ga-Zn-O/Al$_2$O$_3$ thin-film transistors on ultrasmooth and clear nanopaper. *ACS Appl. Mater. Interfaces* **2017**, *9*, 27792–27800. [CrossRef] [PubMed]

13. Kim, J.; Jang, I.; Jeong, J. Effects of helium annealing in low-temperature and solution-processed amorphous indium-gallium-zinc-oxide thin-film transistors. *AIP Adv.* **2019**, *9*, 045228. [CrossRef]

14. Minari, T.; Kanehara, Y.; Liu, C.; Sakamoto, K.; Yasuda, T.; Yaguchi, A.; Tsukada, S.; Kashizaki, K.; Kanehara, M. Room-temperature printing of organic thin-film transistors with π-junction gold nanoparticles. *Adv. Funct. Mater.* **2014**, *24*, 4886–4892. [CrossRef]

15. Choi, C.; Baek, Y.; Lee, B.M.; Kim, K.H.; Rim, Y.S. Enhanced electrical stability of nitrate ligand-based hexaaqua complexes solution-processed ultrathin a-IGZO transistors. *J. Phys. D Appl. Phys.* **2017**, *50*, 485107. [CrossRef]

16. Kim, Y.R.; Kwon, J.H.; Vincent, P.; Kim, D.K.; Jeong, H.S.; Hahn, J.; Bae, J.H. Effect of UV and water on electrical properties at pre- and post-annealing processes in solution-processed inGaZnO transistors. *J. Nanosci. Nanotechnol.* **2019**, *19*, 2240–2246. [CrossRef] [PubMed]

17. Cho, M.H.; Seol, H.; Song, A.; Choi, S.; Song, Y.; Yun, P.S.; Chung, K.-B.; Bae, J.U.; Park, K.S.; Jeong, J.K. Comparative study on performance of igzo transistors with sputtered and atomic layer deposited channel layer. *IEEE Trans. Electron Devices* **2019**, *66*, 1783–1788. [CrossRef]

18. Xu, W.; Xu, M.; Jiang, J.; Xu, S.; Feng, X. Impact of sputtering power on amorphous in-al-zn-o films and thin film transistors prepared by rf magnetron sputtering. *IEEE Trans. Electron Devices* **2019**, *66*, 2219–2223. [CrossRef]

19. Fuh, C.S.; Liu, P.T.; Teng, L.F.; Huang, S.W.; Lee, Y.J.; Shieh, H.P.D.; Sze, S.M. Effects of microwave annealing on nitrogenated amorphous in-ga-zn-o thin-film transistor for low thermal budget process application. *IEEE Electron Device Lett.* **2013**, *34*, 1157–1159. [CrossRef]

20. Lee, Y.S.; Yen, T.W.; Lin, C.I.; Lin, H.C.; Yeh, Y. Electrical characteristics of amorphous in-ga-zn-o thin-film transistors prepared by radio frequency magnetron sputtering with varying oxygen flows. *Displays* **2014**, *35*, 165–170. [CrossRef]

21. Hung, M.P.; Chare, C.; Nag, M.; de Meux, A.D.J.; Genoe, J.; Steudel, S. Effect of high oxygen partial pressure on carrier transport mechanism in a-ingazno tfts. *IEEE Trans. Electron Devices* **2018**, *65*, 2833–2837. [CrossRef]

22. Moon, Y.K.; Lee, S.; Kim, D.H.; Lee, D.H.; Jeong, C.O.; Park, J.W. Application of dc magnetron sputtering to deposition of ingazno films for thin film transistor devices. *Jpn. J. Appl. Phys.* **2009**, *48*, 031301. [CrossRef]

23. Zhang, H.; Wang, Y.; Zhang, X.; Liu, C. Improvement of electrical characteristics and stability of IGZO TFT through surface single crystallization of IGZO film at room temperature. *Semicond. Sci. Technol.* **2018**, *33*, 085015. [CrossRef]

Article

Improving the Tribological Performance of MAO Coatings by Using a Stable Sol Electrolyte Mixed with Cellulose Additive

Wei Song [1,2], Bailing Jiang [1,*] and Dongdong Ji [1]

[1]   Faculty of Materials Science and Engineering, XI'AN University of Technology, NO.5 South Jinhua Road, Xi'an 710048, Shaanxi, China; song78wei@163.com (W.S.); jdd3141592654@163.com (D.J.)

[2]   School of biological and Chemical Engineering, Nanyang Institute of Technology, NO.80 Changjiang Road, Nanyang 473004, He'nan, China

*   Correspondence: jiangbail@vip.163.com

Received: 23 November 2019; Accepted: 12 December 2019; Published: 16 December 2019

Abstract: In this study, micro-arc oxidation (MAO) of aluminum 6061 alloy was carried out within a silicate base electrolyte containing 0.75 g/L of cellulose, and the tribological properties of the coating were investigated. The as-prepared coating was detected by Fourier Transform Infrared Spectroscopy (FTIR), X-ray diffraction (XRD), a scanning electron microscope (SEM) and an energy-dispersive spectrometer (EDS), respectively. The results suggested that cellulose filled in the microcracks and micropores, or it existed by cross-linking with Al^{3+}. In addition, it was found that the cellulose had little effect on the coating hardness. However, the thickness and roughness of the coating were improved with the increase in cellulose concentration. Moreover, the ball-on-disk test showed that the friction coefficient, weight loss and wear rate of the MAO coating decreased with the increase in cellulose concentration. Further, the performances of the coatings obtained in the same electrolyte, under different preserved storage periods, were compared, revealing that the cellulose was uniformly dispersed in the electrolyte and improved the tribological properties of the MAO coating within 30 days.

Keywords: cellulose; tribological performance; stability; MAO (micro-arc oxidation) coating; self-lubricating

---

1. Introduction

Aluminum alloys are characterized by their excellent castability, high specific strength and low thermal expansion coefficient [1]. As a result, they have aroused increasing interest in the automobile industry, as well as having aerospace structural and military applications [2–4]. Nonetheless, aluminum alloys are associated with poor tribological performances, since the friction coefficient is as high as 0.5–0.8 under dry friction conditions [5]. Typically, aluminum alloys exhibit poor tribological performances when they come into contact with other metal materials. This is ascribed to severe adhesive wear, plastic deformation and metallic wear [6]. Consequently, surface modification approaches are indispensable when it comes to enhancing the tribological performances of aluminum alloys.

A variety of surface treatment techniques are available at present, such as the electrochemical approach [7], electroless deposition [8], chemical surface conversion [9], deposition from the gas-phase [10], laser surface alloying [11] and organic polymer coating [12]. Micro-arc oxidation (MAO), also referred to as plasma electrolytic oxidation, sparks anodization or micro-plasma discharge oxidation [13,14] and emerges as a unique technique to produce hard and thick ceramic oxide coatings on diverse Al [15], Mg [16] and Ti [17] alloys. Noteworthily, coatings synthesized according to the MAO process exhibit superior mechanical properties, including excellent adhesive strength [18], high micro-hardness [19], and high thermal conductivity [20] compared with those obtained through other methods. MAO coatings offer several advantages over other coatings. MAO coatings are very stable and hard, which means they can be used at high temperatures. MAO treatment can significantly enhance the surface properties of Mg, Al, Ti and their alloys. For instance, MAO coatings exhibit better anti-wear and anti-corrosion performances than other chemical conversion layers. In addition, the pores and cracks generated in MAO coatings during micro-arc discharges can help relieve the residual stress of the coating. Thus, MAO coatings are promising for the corrosion protection of aluminum and magnesium [21], the wear resistance of light metals and their alloys [22], and the improved biofunctionality of titanium [23]. In addition, the composition, structure, and properties of coatings produced by the MAO process depend on various parameters, among which chemical composition and electrolyte concentration are the most important [24–26]. The microstructures and properties of diverse composites have also been extensively investigated in plenty of reviews and books [27–30].

Generally, composites are added into the electrolyte to improve the tribological performances of aluminum alloys, since they are able to compact the coating by filling in the microcracks and micropores of the MAO coating [31,32], sealing the surface or reacting with the aluminum ion as the coating forming matter [33,34].

However, these techniques are linked with certain shortcomings [35,36]:

1. The carbide and nitride oxide that must be mixed into the metal matrices are so hard and brittle that they may be broken in the course of mixing or in the consolidation processes.
2. The additive may not be uniformly dispersed into the electrolyte.
3. A chemical reaction between the metal matrix and the coating may occur during the exposure to elevating temperature, which leads to poor mechanical properties of the composites.
4. The particle sizes are typically in tens to hundreds of microns, which considerably reduces the ductility and toughness, as well as ineffectively utilizing the strength and stiffness of the reinforcement.
5. The electrolyte is unstable and cannot be used in actual industrial production.

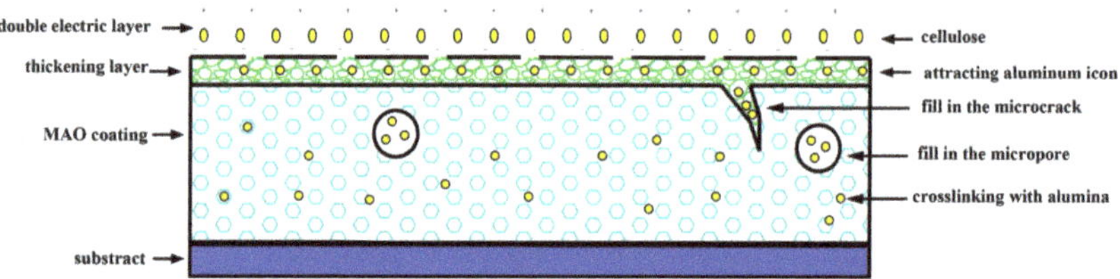

Figure 1. The anti-friction mechanism of cellulose improves the micro-arc oxidation (MAO) coating.

Cellulose is different from the above components because it is a polymer compound that contains multiple hydroxyl groups and experiences limited swelling under alkaline conditions, resulting in the formation of a stable uniform sol electrolyte [37].

Figure 1 illustrates that the anti-friction mechanism of cellulose improves the MAO coating. As a polymer, cellulose possesses favorable self-lubricity and plastic deformation ability. The possible mechanisms by which cellulose typically prompts the tribological performance are explained below.

First, the MAO coating can be decreased depending on the self-lubrication ability of the cellulose. Secondly, the propagation of microcracks and micropores generated by thermal stress in the coating can be inhibited based on the plastic deformation capacity. Thirdly, the cellulose fills in the microcracks and micropores. As a result, the coating compactness is increased when it forms the complex with the aluminum ion.

It is well known that the coating's tribological performance is enhanced with an increase in compactness and a decrease in friction coefficient, whereas an increase in coating toughness [38] reduces the occurrence of adhesive wear [39], which seriously affects the service life of the coating [40].

On this account, mixing cellulose into the electrolyte contributes to obtaining a stable sol solution with uniform dispersion, which is helpful for preparing a MAO coating with excellent tribological performance.

This study mainly aimed to improve the tribological performance of the MAO coating by adding cellulose into the electrolyte. Moreover, the effect of cellulose content on the tribological performance of the coating was investigated under optimized parameters; subsequently, the tribological performances of coatings obtained during different storage periods were compared in order to investigate the stability of the sol electrolyte.

2. Experimental Procedure

2.1. Materials

Aluminum 6061 alloy (AA 6061) specimens with the dimensions of $20 \times 20 \times 3.5$ mm^3 were used as the anodized substrates. Specifically, the alloy composition by wt % included 0.8–1.2% Mg, 0.4–0.8% Si, 0.15–0.4% Cu, and 0.04–0.35% Cr, and Al was the balance. The contents of Fe, Mn, Zn and Ti in the alloy were not higher than 0.7, 0.15, 0.25 and 0.15 wt %, respectively. Prior to the experiment, all specimens were mechanically ground using 240, 400 and 800 grit silicon carbide paper and then washed with distilled water.

2.2. Experiment Process

Two 300×300 mm^2 AISI 321 stainless steel sheets (BENLAIMETAL, Shanghai, China) were used as the cathode. Then, the MAO process was carried out in a stirred electrolyte consisting of 15 g/L Na_2SiO_3 (MACKLIN, Beijing, China), 5 g/L KOH (MACKLIN, Beijing, China), and 5 g/L $(NaPO_3)_6$ (MACKLIN, Beijing, China). Except for the adjusted cellulose concentrations (0, 0.25, 0.50, 0.75, and 1 g/L), the MAO processes of specimens were carried out at 20 °C for 30 min with a DC pulse supply at the frequency of 50 kHz, the constant current density of 1 A/cm^2, and the duty cycle of 15%. To examine the electrolyte stability, the MAO processes were conducted within the same electrolyte after various storage periods for 0, 1, 7, 14 and 30 days.

2.3. Characterization

The phase compositions of diverse coatings were examined through an X-ray diffractometer (XRD) (D/max-rB, RICOH, Tokyo, Japan) with a Cu Ka source, and the accelerating voltage and applied current were 40 kV and 30 mA, respectively. In addition, the radiation emitted by the sample surface was detected by a Fourier transform infrared (FT-IR) spectrometer (JASCO FT/IR-6100, JASCO, Toykp, Japan). Meanwhile, scanning electron microscopy (SEM, JSM-6700F, JEOL, Japan) was employed to observe the microstructure and morphology of the MAO coatings, whereas the element compositions on the coating were analyzed by an energy-dispersive spectrometer (EDS, Oxford, UK) combined with SEM. Further, the coating thicknesses generated under different conditions were measured using an eddy current coating thickness measurement gauge (CTG-10, Time Company, Beijing, China). To be specific, the thicknesses at 10 different sites on the coating surface were measured to calculate and record the average. Additionally, the coating roughness was tested using a roughness tester (TR-3200, Time Company, Beijing, China, vertical resolution of 0.01 μm). The micro-hardness of the coating was evaluated using the HVS-100 micro-hardness tester (TMVS-1, TIMES Group, Beijing, China) with a load of 100 g for 10 s. In addition, the tribological behaviors of the coatings were evaluated using the ball-on-disk tester (UMT-Tribolab, BRUKER, Bremen, Germany) under dry sliding conditions. Typically, balls of GCr15 with a diameter of 10 mm and a hardness of HRC 60 were used as the counterface materials. The normal load was 5 N, and the linear sliding speed was 0.01 m/s. All tests were run under the laboratory conditions (temperature of 25 °C and relative humidity of 50%) for 30 min each. The friction coefficient was recorded on a computer during each test. The wear loss was weighed using an electronic balance, and the wear rate (k) was calculated according to the following Formula (1).

$$k = \frac{\pi \bullet D[\arcsin(\frac{L_u}{2r})r^2 - L_u(\frac{\sqrt{4r^2 - L_u^2}}{4})]}{P \times S} \tag{1}$$

where r stands for the radius of the corundum ball (mm), D represents the diameter of the wear track (mm), L_u indicates the width of the wear track (mm), S is the sliding distance (m), and P is the applied normal load (N).

3. Results and Discussion

3.1. Thickness and Roughness of the MAO Coating

The tribological performance of the MAO coating was affected by its thickness and roughness; therefore, the impacts of cellulose content on the coating thickness and roughness were examined, as shown in Figure 2.

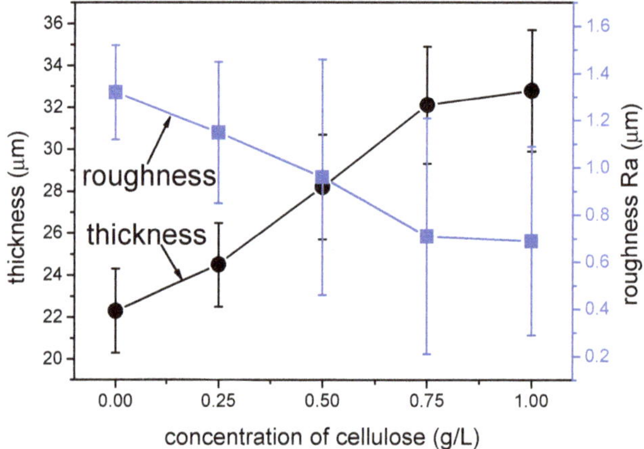

Figure 2. Effects of cellulose content on the thickness and roughness of the MAO coating.

It can be observed from Figure 2 that with the increase in cellulose concentration, the thickness of the MAO coating increased, whereas its roughness decreased. At the cellulose content of 0.75 g/L, the coating thickness and roughness reached 32.1 µm microns and 0.66 µm, respectively. However, further increase in the cellulose concentration showed no obvious improvement in the thickness and roughness of the MAO coating, which was mainly ascribed to the low cellulose content. Specifically, it contained the multiple-hydroxy, and a double electric layer was formed during the electrochemical process, which attracted Al^{3+} contiguous to the substrate surface [38], resulting in the increase in the MAO coating thickness. In addition, cellulose participated in the coating formation by filling in the microcracks and micropores, even cross-linking with Al^{3+} in the coating, as observed in Figure 1. Moreover, the excellent polymer plasticity contributed to reducing the quantity and size of microcracks and micropores, thus decreasing the MAO coating roughness. Nonetheless, the Al^{3+} escaping from the substrate was limited by the electrochemical parameters, and further increase in the cellulose contents showed no significant improvement of the coating thickness and roughness when most of the Al^{3+} ions were attracted by the cellulose. Additionally, the electrolyte became inhomogeneous after over 24 h of preservation, so the optimal cellulose content was determined to be 0.75 g/L.

3.2. Microstructure of the MAO Coating

The surfaces and cross-section microstructures of the MAO coatings at different cellulose contents were observed through SEM. The results are shown in Figures 3 and 4, respectively. As observed from Figure 3, the increase in cellulose content led to the decreased size of the microcracks and micropores, while it increased the quantity of the micropores.

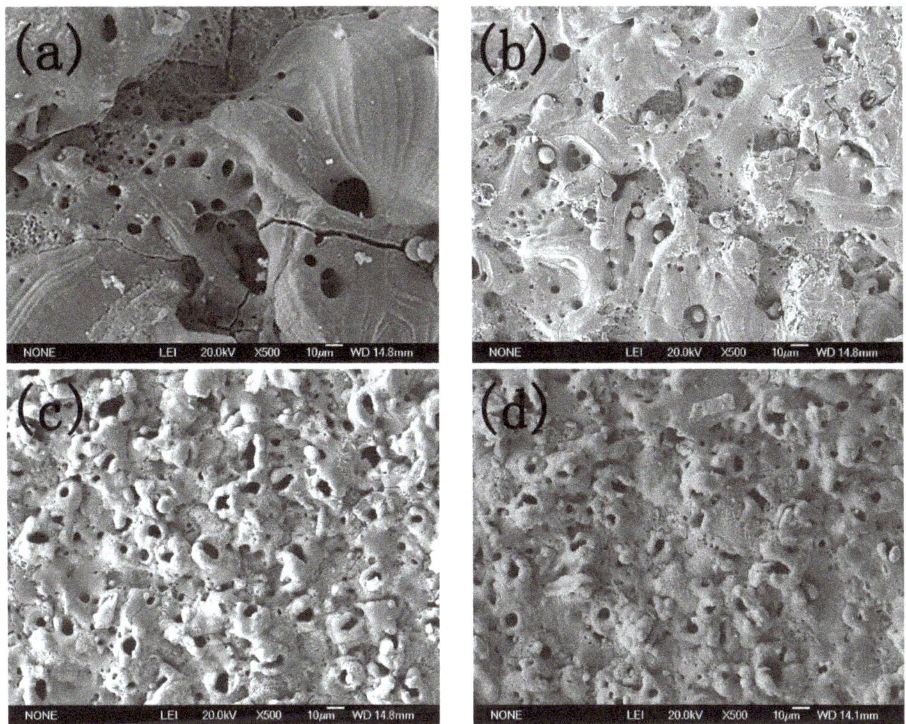

Figure 3. Microsurface of the MAO coating at cellulose concentrations of (**a**) 0 g/L, (**b**) 0.25 g/L, (**c**) 0.50 g/L, and (**d**) 0.75 g/L.

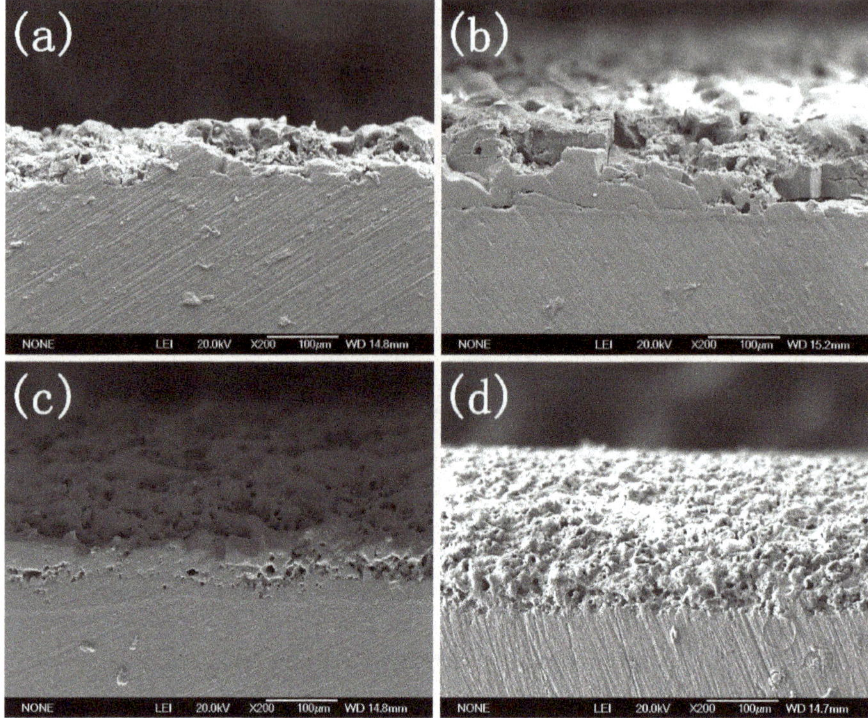

Figure 4. Cross-section of the MAO coating at cellulose concentrations of (**a**) 0 g/L, (**b**) 0.25 g/L, (**c**) 0.50 g/L, and (**d**) 0.75 g/L.

The cross-section photograph displayed in Figure 4 proves the above findings. In addition, Figure 4 also suggests that when the cellulose concentration was 0 g/L, the coating thickness was small and the adhesion between coating and substrate was poor (seen in Figure 4a). With the increase in cellulose concentration, the coating thickness increased and the adhesion between coating and substrate was enhanced (seen in Figure 3b,c,d). The possibility of adhesive wear was reduced with the increase in the bonding force between the coating and the substrate. To further investigate the coating component, EDS was carried out. The carbon contents at different sites are presented in Figure 5 and Table 1. As displayed in Table 1, the carbon element spread all over the coating, which proved that part of the cellulose filled in the microcracks and micropores, while part of it cross-linked with the Al^{3+} in the coating. In addition, the cellulose content in the micropores and microcracks was higher than it was at the other sites, indicating that they were filled in by a relatively small portion of the cellulose.

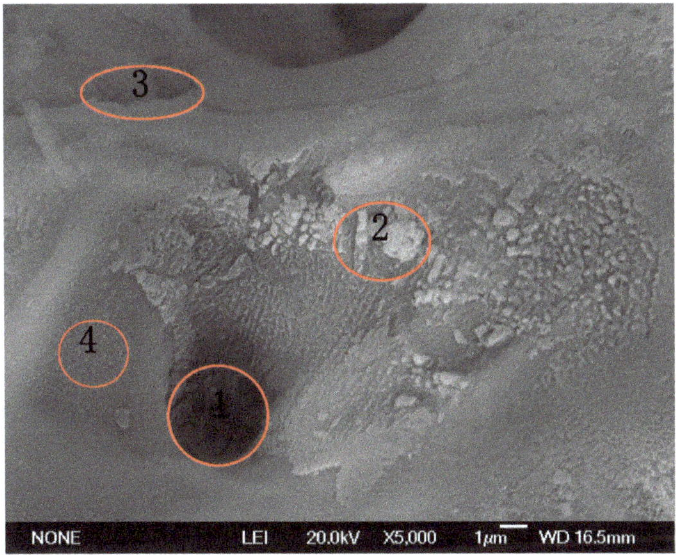

Figure 5. Micrograph illustrating the zone of energy-dispersive spectrometer (EDS) analysis.

Table 1. C, O, Al element contents at different positions.

Point	C K (at. %)	O K (at. %)	Al K (at. %)
1	4.21	45.31	42.48
2	3.16	48.43	45.41
3	3.83	46.54	43.63
4	2.43	49.84	46.73

3.3. Phase Structure of the Coating

The crystalline phase compositions of the MAO coatings at the cellulose contents of 0, 0.25, 0.50, 0.75 and 1 g/L were analyzed by means of FTIR and XRD, respectively. The results are shown in Figures 6 and 7, separately. Figure 6 illustrates the infrared absorption peaks of the MAO coatings obtained under various cellulose contents. Notably, the peaks at 3406 cm^{-1} were assigned to O–H stretching vibrations, while those at 1630 cm^{-1} corresponded to C–O stretching vibrations, and those at 838 and 648 cm^{-1} were indexed to Al–O stretching vibrations. As indicated by Figure 7, the increase in the cellulose content gave rise to the enhanced characteristic peak of the cellulose and the decreased peak intensity of the alumina. Taken together, the analyzed results of the FTIR and XRD spectra proved the presence of cellulose in the MAO coating.

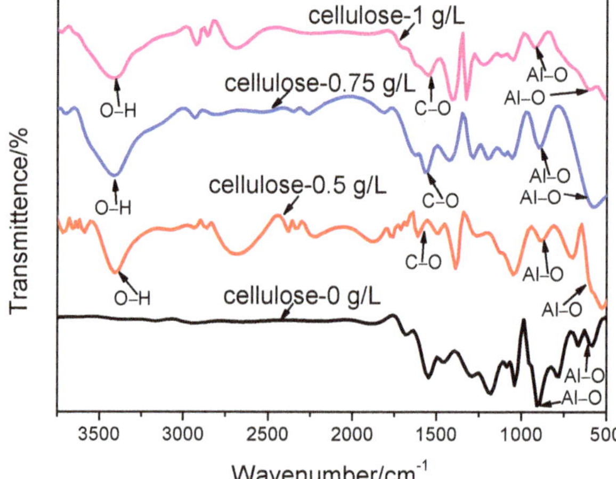

Figure 6. Fourier transform infrared (FT-IR) spectra of the MAO coatings obtained at different cellulose contents.

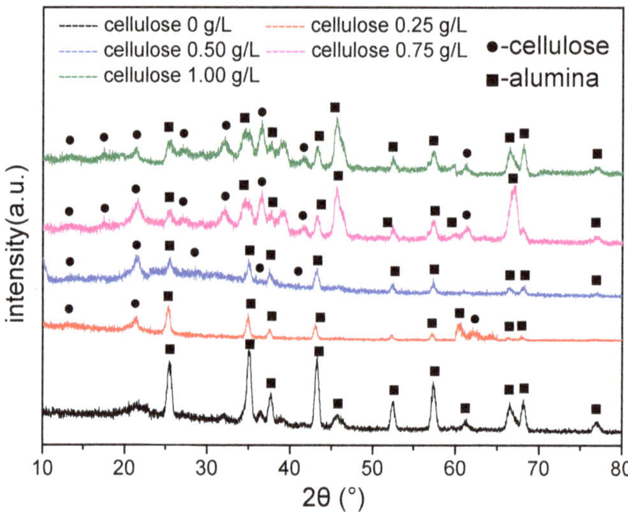

Figure 7. X-ray diffraction (XRD) spectra of the MAO coatings obtained at different cellulose contents.

3.4. Tribological Performances of the MAO Coatings

Figure 8 exhibits the influences of the cellulose content on the friction coefficient. Clearly, the friction coefficient was significantly reduced after the aluminum alloys were treated by the MAO technology, and it slowly decreased with the further increase in the cellulose content. In addition, the friction coefficients of most samples were maintained at fixed values when cellulose was used as the additive; however, that of the cellulose-free MAO coating was suddenly increased after 20 min. These findings revealed that the MAO surface treatment technology reduced the friction coefficient of the aluminum alloy, and the addition of cellulose into the electrolyte was beneficial to further decrease and maintain the friction coefficient for a long time. To examine the tribological properties of the MAO coatings obtained at different cellulose contents, the micro-hardness of the MAO coatings were tested by the micro-hardness tester, while the wear loss (the amount of material lost during the mechanical tests) and wear rate were determined through wear tests. Results are presented in Table 2.

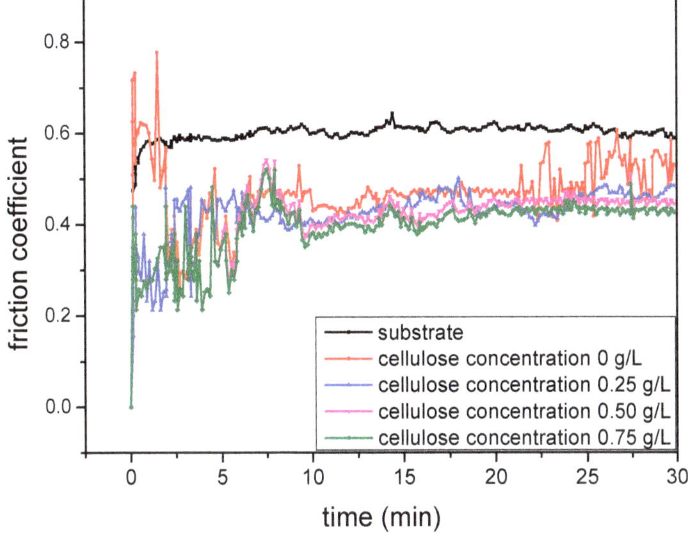

Figure 8. Friction coefficients of the MAO coatings obtained at different cellulose contents.

Table 2. Hardness and tribological performances of the MAO coatings obtained at different cellulose contents.

Cellulose Concentration g/L	Micro-Hardness $HV_{0.1}$	Weight Loss mg	Wear Track Depth μm	Wear Track Width μm	Wear Rate 10^{-5} $mm^3/N\cdot m$
0	1260	16	6.42	679.61	2.30
0.25	1230	14	4.59	537.49	1.13
0.50	1240	12	4.31	522.32	1.03
0.75	1230	11	3.54	486.26	0.84
1.00	1220	10	3.13	482.01	0.82

According to Table 2, the micro-hardness of the coatings remained at about 1230 $HV_{0.1}$, while the wear loss and wear rate decreased when the cellulose content was elevated from 0% to 1.0%. These results suggest that the addition of cellulose was beneficial for improving the tribological performances of the MAO coatings.

3.5. Stability of the Electrolyte

Apart from the favorable anti-wear performance, the electrolyte stability, especially when polymer is used as the additive, is also a crustal parameter in practical industrial production. To investigate the electrolyte stability during long-term storage, the performances of the MAO coatings (such as thickness, roughness, hardness, friction coefficient, wear loss and wear rate) obtained at different electrolyte storage periods were compared, as shown in Figure 9 and Table 3. There was no obvious difference between them, demonstrating that the electrolyte might be employed to improve the tribological performance of the MAO coating within 30 days.

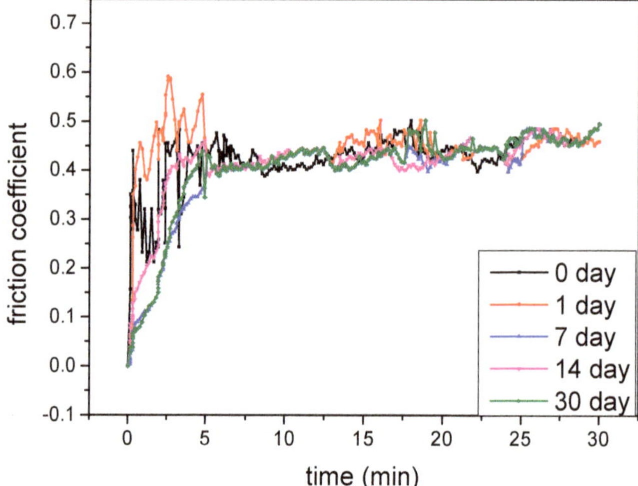

Figure 9. Friction coefficients of the MAO coatings under different storage periods.

Table 3. Coating performances under different storage periods.

Stability of the Electrolyte Day	Micro-Hardness $HV_{0.1}$	Weight Loss mg	Thickness μm	Roughness μm	Wear Rate 10^{-5} $mm^3/N \cdot m$
0	1230	11	32.1	0.66	0.84
1	1230	15	32.2	0.72	0.79
7	1220	10	31.9	0.68	0.91
14	1230	11	32.1	0.69	0.89
30	1220	11	31.9	0.67	0.92

4. Conclusions

1. The tribological performances of MAO coatings are improved by mixing 0.75 g/L of cellulose into the electrolyte. The thickness of the coating increases while the roughness decreases with the increase in cellulose content.

2. The coating compositions are thereby analyzed by FTIR and XRD, which prove the presence of cellulose in the coating. Moreover, the coating microstructure is observed through SEM, which reveals that the coating has a compact structure; meanwhile, the coating compositions at micropores, microcracks and normal positions are examined through EDS, suggesting that part of the cellulose fills in the microcracks and micropores, and part of it cross-links with the Al^{3+}.

3. The tribological performances of the coatings at different cellulose concentrations are evaluated using a ball-on-disk tester under dry sliding conditions. After different storage periods, they are compared at the same electrolyte, revealing that MAO coatings with consistent quality can be produced in this electrolyte.

Author Contributions: Conceptualization, W.S., B.J., D.J.; methodology, W.S., B.J.; writing—original draft preparation, W.S.; writing—review and editing, D.J.

Funding: This research received no external funding.

Acknowledgments: This work was supported by the National Natural Science Foundation of China (Grant No. 51571114).

References

1. Jeon, G.T.K.; Kim, Y.; Moon, J.H.; Lee, C.; Kim, W.J.; Kim, S.J. Effect of Al 6061 Alloy Compositions on Mechanical Properties of the Automotive Steering Knuckle Made by Novel Casting Process. *Metals* **2018**, *8*, 857. [CrossRef]
2. Liu, Z.Y.; Xiong, B.Q.; Li, X.W.; Yan, L.Z.; Li, Z.H.; Zhang, Y.A.; Liu, H.W. Deep drawing of 6A16 aluminum alloy for automobile body with various blank-holder forces. *Rare Met.* **2019**, *38*, 946–953. [CrossRef]
3. Kovalov, D.; Fekete, B.; Engelhardt, G.R.; Macdonal, D.D. Prediction of corrosion fatigue crack growth rate in alloys. Part I: General corrosion fatigue model for aero-space aluminum alloys. *Corros. Sci.* **2018**, *141*, 22–29. [CrossRef]
4. Lenihan, D.; Ronan, W.; O'Donoghue, P.E.; Leen, S.B. A review of the integrity of metallic vehicle armour to projectile attack. *Proc. Inst. Mech. Eng. Part L-J. Mater.* **2019**, *233*, 73–94. [CrossRef]
5. Benea, L.; Dumitrascu, V. Enhancement in sustained friction and wear resistance of nanoporous aluminum oxide films obtained by controlled electrochemical oxidation process. *Rsc. Adv.* **2019**, *9*, 25056–25063. [CrossRef]
6. Sardar, S.; Pradhan, S.K.; Karmakar, S.K.; Das, D. Modeling of Abraded Surface Roughness and Wear Resistance of Aluminum Matrix Composites. *J. Tribol.* **2019**, *141*, 071601. [CrossRef]
7. Ghafaripoor, M.; Raeissi, K.; Santamaria, M.; Hakimizad, A. The corrosion and tribocorrosion resistance of PEO composite coatings containing alpha-Al2O3 particles on 7075 Al alloy. *Surf. Coat. Tech.* **2018**, *349*, 470–479. [CrossRef]
8. Stark, L.M.; Smid, I.; Segall, A.E.; Eden, T.J.; Potter, J. Self-Lubricating Cold-Sprayed Coatings Utilizing Microscale Nickel-Encapsulated Hexagonal Boron Nitride. *Tribol. Trans.* **2012**, *55*, 624–630. [CrossRef]
9. Wu, L.; Ding, X.X.; Zheng, Z.C.; Ma, Y.L.; Andrej, A.; Chen, X.B.; Xie, Z.H.; Su, D.E.; Pan, F.S. Fabrication and characterization of an actively protective Mg-Al LDHs/Al2O3 composite coating on magnesium alloy AZ31. *Appl. Surf. Sci.* **2019**, *487*, 558–568. [CrossRef]
10. Blomqvist, A.; Århammar, C.; Pedersen, H.; Silvearv, F.; Norgren, S.; Ahuja, R. Understanding the catalytic effects of H2S on CVD-growth of alpha-alumina: Thermodynamic gas-phase simulations and density functional theory. *Surf. Coat. Tech.* **2011**, *206*, 1771–1779. [CrossRef]
11. SINGH, A.; HARIMKAR, S.P. Laser Surface Engineering of Magnesium Alloys: A Review. *JOM-US.* **2012**, *64*, 716–733. [CrossRef]
12. Chen, J.M.; Yang, J.; Zhao, X.Q.; An, Y.L.; Hou, G.L.; Chen, J.; Zhou, H.D. Preparation and properties of poly-(p)-oxybenzoyl/aluminum bronze composite coating by atmosphere plasma spraying. *Surf. Coat. Tech.* **2014**, *253*, 261–267. [CrossRef]
13. Gu, Y.H.; Ma, H.J.; Yue, W.; Tian, B.; Chen, L.L.; Mao, D.L. Microstructure and corrosion model of MAO coating on nano grained AA2024 pretreated by ultrasonic cold forging technology. *J. Alloy. Compd.* **2016**, *681*, 120–127. [CrossRef]
14. Tonelli, L.; Pezzato, L.; Dolcet, P.; Dabalà, M.; Martini, C. Effects of graphite nano-particle additions on dry sliding behaviour of plasma-electrolytic-oxidation-treated EV31A magnesium alloy against steel in air. *Wear* **2018**, *404*, 122–132. [CrossRef]
15. Yi, P.; Yue, W.; Liang, J.; Hou, B.B.; Sun, J.H.; Gu, Y.H.; Liu, J.X. Effects of nanocrystallized layer on the tribological properties of micro-arc oxidation coatings on 2618 aluminum alloy under high temperatures. *Int. J. Adv. Manuf. Tech.* **2018**, *96*, 1635–1646. [CrossRef]
16. Zhang, D.Y.; Ge, Y.F.; Liu, G.L.; Gao, F.; Li, P.Y. Investigation of tribological properties of micro-arc oxidation ceramic coating on Mg alloy under dry sliding condition. *Ceram. Int.* **2018**, *44*, 16164–16172. [CrossRef]
17. Demirbas, C.; Ayday, A. The influence of Nano-TiO2 and Nano-Al2 O3 Particles in Silicate Based Electrolytes on Microstructure and Mechanical Properties of Micro Arc Coated Ti6Al4V Alloy. *Mater. Res.* **2018**, *21*, e20180092. [CrossRef]
18. Dudarev, N.; Gallyamova, R. The Cnfluence of Chemical Composition of Aluminum Alloys on the Quality of Oxide Layers Formed by Microarc Oxidation. *Mater. Today-Proc.* **2019**, *11*, 89–94. [CrossRef]
19. Tran, Q.P.; Kuo, Y.C.; Sun, J.K.; He, J.L.; Chin, T.S. High quality oxide-layers on Al-alloy by micro-arc oxidation using hybrid voltages. *Surf. Coat. Tech.* **2016**, *303*, 61–67. [CrossRef]

20. Tang, H.; Sun, Q.; Xin, T.Z.; Yi, C.G.; Jiang, Z.H.; Wang, F.P. Influence of Co(CH3COO)(2) concentration on thermal emissivity of coatings formed on titanium alloy by micro-arc oxidation. *Curr. Appl. Phys.* **2012**, *12*, 284–290. [CrossRef]

21. Muhafffel, F.; Cimenoglu, H. Development of corrosion and wear resistant micro arc oxidation coating on a magnesium alloy. *Surf. Coat. Tech.* **2019**, *357*, 822–832. [CrossRef]

22. Dehghanghadikolaei, A.; Ibrahim, H.; Amerinatanzi, A.; Hashemi, M.; Moghaddam, N.S.; Elahinia, M. Improving corrosion resistance of additively manufactured nickel–titanium biomedical devices by micro arc oxidation process. *J. Mater. Sci.* **2019**, *54*, 7333–7355. [CrossRef]

23. Sobolev, A.; Wolicki, I.; Kossenko, A.; Zinigrad, M.; Borodianskiy, K. Coating Formation on Ti-6Al-4V Alloy by Micro Arc Oxidation in Molten Salt. *Materials* **2018**, *11*, 1611. [CrossRef] [PubMed]

24. Tong, Y.G.; Zhou, Z.B.; Cai, H.; Wang, X.L.; Wang, Y. Preparation and Properties of a Flexible Al2O3/Al/Al2O3 Composite. *Adv. Mater. Sci. Eng.* **2018**, *2018*, 4739267. [CrossRef]

25. Chen, Q.Z.; Jiang, Z.Q.; Tang, S.G.; Dong, W.B.; Tong, Q.; Li, W.Z. Influence of graphene particles on the micro-arc oxidation behaviors of 6063 aluminum alloy and the coating properties. *Appl. Surf. Sci.* **2017**, *423*, 939–950. [CrossRef]

26. Balaji, R.; Pushpavanam, M.; Kumar, K.Y.; Subramanian, K. Electrodeposition of bronze-PTFE composite coatings and study on their tribological characteristics. *Surf. Coat. Tech.* **2006**, *201*, 3205–3211. [CrossRef]

27. Atiyeh, B.S.; Costagliola, M.; Hayek, S.N.; Dibo, S.A. Effect of silver on burn wound infection control and healing: review of the literature. *Burn* **2007**, *33*, 139–148. [CrossRef]

28. Burg, K.J.L.; Porter, S.; Kellam, J.F. Biomaterial developments for bone tissue engineering. *Biomaterials* **2000**, *21*, 2347–2359. [CrossRef]

29. Fukumasa, O.; Tagashira, R.; Tachino, K.; Mukunoki, H. Spraying of MgO fifilms with a well-controlled plasma jet. *Surf. Coat. Tech.* **2003**, *169*, 579–582. [CrossRef]

30. Li, Z.W.; Di, S.C. Effect of Anode Pulse-Width on the Microstructure and Wear Resistance of Microarc Oxidation Coatings. *Metals* **2017**, *7*, 243. [CrossRef]

31. Li, H.; Sun, Y.Z.; Zhang, J. Effect of ZrO2 particle on the performance of micro-arc oxidation coatings on Ti6Al4V. *Appl. Surf. Sci.* **2015**, *342*, 183–190. [CrossRef]

32. Lin, X.Z.; Zhu, M.H.; Cai, Z.B.; Zheng, J.F.; Peng, J.F. Microstructure and Reciprocating Sliding Tribological Performance of Micro-arc Oxidation Coating Prepared on Al-Si Alloy. *Adv. Mater. Res.* **2010**, *97–101*, 1518–1526. [CrossRef]

33. Tsunekawa, S.; Aoki, Y.; Habazaki, H. Two-step plasma electrolytic oxidation of Ti–15V–3Al–3Cr–3Sn for wear-resistant and adhesive coating. *Surf. Coat. Tech.* **2011**, *205*, 4732–4740. [CrossRef]

34. Li, X.; Luan, B.L. Discovery of Al2O3 particles incorporation mechanism in plasma electrolytic oxidation of AM60B magnesium alloy. *Mater. Lett.* **2012**, *86*, 88–91. [CrossRef]

35. Yao, Z.; Jiang, Y.; Jiang, Z.; Wang, F.; Wu, Z. Preparation and structure of ceramic coatings containing zirconium oxide on Ti alloy by plasma electrolytic oxidation. *J. Mater. Process. Tech.* **2008**, *205*, 303–307. [CrossRef]

36. Huang, S.Q.; Wu, L.J.; Li, T.Z.; Xu, D.Y.; Lin, X.L.; Wu, C.D. Facile preparation of biomass lignin-based hydroxyethyl cellulose superabsorbent hydrogel for dye pollutant removal. *Int. J. Biol. Macromol.* **2019**, *137*, 939–947. [CrossRef] [PubMed]

37. Wei, T.B.; Guo, B.G.; Yan, F.Y.; Tian, J.; Liang, J. Tribological Properties of Micro-arc Oxidation Coatings on Al Alloy. *J. Mater. Sci. Eng.* **2004**, *22*, 564–567. [CrossRef]

38. Liu, F.; Xu, J.L.; Yu, D.Z.; Wang, F.P.; Zhao, L.C. Wear resistance of micro-arc oxidation coatings on biomedical NiTi alloy. *J. Alloy. Compd.* **2009**, *487*, 391–394. [CrossRef]

39. Zhu, M.H.; Cai, Z.B.; Lin, X.Z.; Zheng, J.F.; Luo, J.; Zhou, Z.R. Fretting wear behaviors of micro-arc oxidation coating sealed by grease. *Wear* **2009**, *267*, 299–307. [CrossRef]

40. Gao, D.D.; Dou, J.H.; Hu, C.; Yu, H.J.; Chen, C.Z. Corrosion behaviour of micro-arc oxidation coatings on Mg-2Sr prepared in poly (ethylene glycol)-incorporated electrolytes. *Rsc. Adv.* **2018**, *8*, 3846–3857. [CrossRef]

Article

A Novel Self-Adaptive Control Method for Plasma Electrolytic Oxidation Processing of Aluminum Alloys

Kai Yang [1,*], Jiaquan Zeng [2], Haisong Huang [1], Jiadui Chen [1] and Biao Cao [2,*]

[1] Key Laboratory of Advanced Manufacturing Technology of the Ministry of Education, Guizhou University, Guiyang 550025, China

[2] College of Mechanical and Automotive Engineering, South China University of Technology, Guangzhou 510640, China

* Correspondence: kyang3@gzu.edu.cn (K.Y.); mebcao@scut.edu.cn (B.C.);
Tel.: +86-159-0201-8391 (K.Y.); +86-138-2504-3678 (B.C.)

Received: 11 August 2019; Accepted: 23 August 2019; Published: 27 August 2019

Abstract: Plasma electrolytic oxidation processing is a novel promising surface modification approach for various materials. However, its large-scale application is still restricted, mainly due to the problem of high energy consumption of the plasma electrolytic oxidation processing. In order to solve this problem, a novel intelligent self-adaptive control technology based on real-time active diagnostics and on the precision adjustment of the process parameters was developed. Both the electrical characteristics of the plasma electrolytic oxidation process and the microstructure of the coating were investigated. During the plasma electrolytic oxidation process, the discharges are maintained in the soft-sparking regime and the coating exhibits a good uniformity and compactness. A total specific energy consumption of 1.8 kW h m^{-2} μm^{-1} was achieved by using such self-adaptive plasma electrolytic oxidation processing on pre-anodized 6061 aluminum alloy samples.

Keywords: plasma electrolytic oxidation; electrical characteristic; anodizing; oxidation; aluminum

1. Introduction

Plasma electrolytic oxidation processing allows the creation of durable, thick, uniform and strongly adherent coatings on valve metals [1,2]. Extensive investigations on plasma electrolytic oxidation have been carried out in past years, due to its increasing use in industrial applications. Most of the research focuses on processing conditions [3–8], discharge characteristics [1,9–13], coating microstructure [14–16], mechanical properties [17], environmental performance [16,18], and functional characteristics [2,18]. As a result, the correlations [3–5,13–15,19] between the electrical conditions, the electrolyte compositions, the coating microstructure, and the growth rate of coating, which are linked via the characteristics of the discharges, have become clearer over recent years. Moreover, progress has been made [20–25] in unraveling the inter-relation between the energetics of individual discharges, the pulse energy, the soft-sparking regime, and the energy consumption.

Recently, the main research focus has been on the dielectric breakdown of oxide films and on the associated discharges that repeatedly occur over the whole surface of the metal substrate. It is clear that most of the new oxide of the coating created during each discharge is formed within the plasma as it cools and collapses [1,5,26]. Although the nature of the plasma created via the discharges, which occur during the plasma electrolytic oxidation process, is still uncertain, it is clear that the discharge characteristics are affected by a series of electrical conditions. Hence, there is considerable scope for more effective process electrical control, with specific objectives in terms of coating performance and energy efficiency, and an attempt is made to identify key points that are likely to assist this.

Several common electrical conditions, such as the use of direct current (DC), alternating current (AC), unipolar, and bipolar pulses, have been employed over recent years. It has been repeatedly found that the AC mode can provide more effective processing and higher quality coatings [1,27]. Furthermore, square waveforms and higher frequencies are being increasingly exploited in research and in commercial use, due to their more sophisticated impulse energy control capabilities [28–30]. However, the most common control modes at the basis of the plasma electrolytic oxidation processing can be classified as current-control and voltage-control. Meanwhile, most of the electrical parameters, which are applied in practice, are fixed or are chosen within a range of preset values on a largely empirical basis [31]. The major disadvantage of such control modes is that the applied electrical parameters cannot be adjusted automatically to match the dynamic requirements of the plasma electrolytic oxidation coating during its growth process.

Furthermore, high energy consumption restricts the performance of traditional plasma electrolytic oxidation processing; by exposing a metal substrate to oxidizing agents and highly energetic discharges, in fact, its discharge mechanism is extremely inefficient. Troughton et al. explored the energies absorbed by various phenomena taking place during a plasma electrolytic oxidation process (melting and vaporization of substrate, melting of existing oxide coating, initiation and sustaining of the plasma, vaporization of water, and electrical heating of the electrolyte) and inferred that most of the injected energy is absorbed in the form of the vaporization of water [32]. What is more, there are several studies [1,24,25] aimed at high energy efficiency and helpful measures. The most promising approach is probably to somehow reduce the energy associated with each discharge, possibly by reducing the voltage needed for it to occur, or to promote more transformation per discharge. Therefore, it is urgent to develop an energy-efficient intelligent process control method.

In this study, a novel self-adaptive control method based on a real-time diagnostic and a precision pulse energy regulation technique was employed for the plasma electrolytic oxidation processing of 6061 aluminum alloys. Moreover, the electrical characteristics, coating microstructure, and energy consumption of such processes were investigated.

2. Materials and Methods

2.1. Materials

A quantity of 6061 aluminum alloy samples with dimensions of 60 mm (L) × 60 mm (W) × 2 mm (H) and a 2 L water-cooled stainless steel tank, which served as a counter-electrode, were used. KOH (5 g L^{-1}) and Na$_2$SiO$_3$ (10 g L^{-1}) were dissolved in distilled water to form the electrolyte. The plasma electrolytic oxidation process was performed by using a home-built 20 kW unipolar pulsed power supply, working in voltage-control mode. The supply could provide a maximum voltage of 600 V and a maximum frequency of 10 kHz. The temperature of the setup was maintained at 25 °C.

2.2. Self-Adaptive Control Method

The applied self-adaptive control model was operated in a double-closed loop (Figure 1a). The voltage loop ensured a constant-voltage control, whereas the voltage–current loop ensured the online adjustment of the process parameters via an active diagnostic method. The complete workflow of the self-adaptive plasma electrolytic oxidation process investigated in this paper is presented in Figure 1b. Initially, the samples were anodized up to 300 V at a rate of 20 V/min. Successively, initial test pulses were applied to determine the initial breakdown voltage (V_i) and the termination voltage (Vu) of the sample. These parameters are listed in Table 1 (type 1). Finally, the plasma electrolytic oxidation processing was implemented by using the self-adaptive unipolar voltage pulsed mode (f = 100 Hz, d = 50%). The applied voltage was dynamically adjusted by using Equation (1):

$$U = V_b + \Delta U. \tag{1}$$

Here, U is the voltage applied to the sample, V_b corresponds to the breakdown voltage, and ΔU represents the voltage deviator (0–5 V).

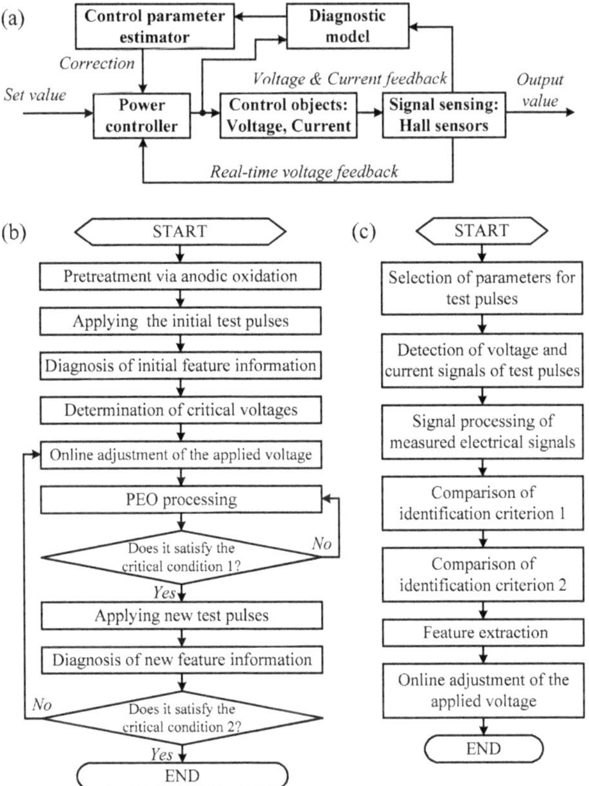

Figure 1. Process control method: (**a**) Self-adaptive control model, (**b**) self-adaptive control process, and (**c**) feature information identification algorithm.

Table 1. Parameters obtained during the test pulses experiment.

Type	Frequency (f, Hz)	Duty Cycle (d)	Number (n)	Growth (ΔP, V)	Basic Value (U_e, V)
1	100	50%	45	8	100
2	100	50%	5 × 5 *	5	V_{bn-1}

* In type 2, the step number of test pulses is equal to 5, the pulse number for the same step is equal to 5, and V_{bn-1} corresponds to the last breakdown voltage.

The processing times to maintain the same applied voltage magnitude and for completing the total process were determined by using the critical condition 1 (Equation (2)) and 2 (Equation (3)), respectively.

$$I \leq \frac{1}{2}I_b. \tag{2}$$

$$U \geq V_u. \tag{3}$$

In Equations (2) and (3), I and U represent the real-time feedback current and the voltage, whereas I_b corresponds to the breakdown current.

The key identification algorithm to obtain the coating feature information is shown in Figure 1c. The parameters of the test pulses acquired during the plasma electrolytic oxidation process are listed in Table 1 (type 2). The identification criterion 1 was used to determine the dynamic values of V_b and I_b as follows:

$$\text{If } \left(\frac{di}{dt}\right)_n \approx 0 \text{ and } \left(\frac{di}{dt}\right)_{n+1} \geq 1, \text{ then } V_b = V_n, \ I_b = I_n.$$

The identification criterion 2 was used to determine the dynamic values of V_u as follows:

$$\text{If } \left(\frac{di}{dt}\right)_n \geq 0, \ \left(\frac{di}{dt}\right)_{n+1} \leq 0, \text{ and } (I_n > I_b), \text{ then } V_u = V_n,$$

where, $\left(\frac{di}{dt}\right)_n$ is the rate of current change of pulse n, $\left(\frac{di}{dt}\right)_{n+1}$ is the rate of current change of pulse $n + 1$, V_n represents the voltage magnitude of pulse n, and I_n represents the current magnitude of pulse n.

A detailed description of the pulse test method and of the feature information extraction can be found in a previous work [6].

2.3. Data Monitoring and Pre-Processing

The voltage and the current data were detected by using a hall voltage sensor (CHV-25P) and a hall current sensor (CSM010B), and recorded with a sampling frequency of 1 MHz via a data acquisition card NI PCI-6133 controlled by LabVIEW software.

Considering the signal interference factors that exist in the plasma electrolytic oxidation process, such as mechanical vibration, electron avalanche, dielectric breakdown, and high-frequency electronic switching, a pre-processing of raw voltage and current data was carried out. The data were first filtered with a low-pass filter with a cut-off frequency of 50 kHz. This allowed high-frequency noise due to electronic switching to be eliminated. Next, these data were smoothed using a moving average filter. Finally, an appropriate scale function and wavelet basis function were adopted to extract a random noise signal.

2.4. Post-Processing of Samples

The surface morphology, the cross-section, and the chemical composition of the coatings prepared on the samples were observed by employing a LEO1530 VP scanning electron microscope (SEM) equipped with an X-ray energy dispersive spectroscopy (EDS) setup. The phase of the coating was analyzed via X-ray diffraction (XRD, Model D8 Advance) by using a Cu kα radiation source.

3. Results

3.1. Electrical Characteristics of the Plasma Electrolytic Oxidation Process

Figure 2a shows that the current waveform changes by following a well-defined trend: When the voltage reached the corrosion potential of the substrate (Figure 2b, about 150 V), the current increased slightly due to the dissolution of the precursor film. Upon the growth of the oxide film, the plasma electrolytic oxidation system tended to stabilize and the current slowly dropped. When the voltage reached 330 V, the current abruptly increased (Figure 2c) and this might result in the dielectric breakdown of the oxide film. At this point, the current increased sharply when the voltage was further increased, implying that higher voltages led to more pronounced discharge events. Figure 2d shows the trend of the oscillating current for voltages higher than 450 V. In these conditions, powerful discharges might occur, causing destructive effects in the sample as reported in previous literature [6,11]. In this work, an initial V_b equal to 330 V and a V_u value of 450 V were chosen.

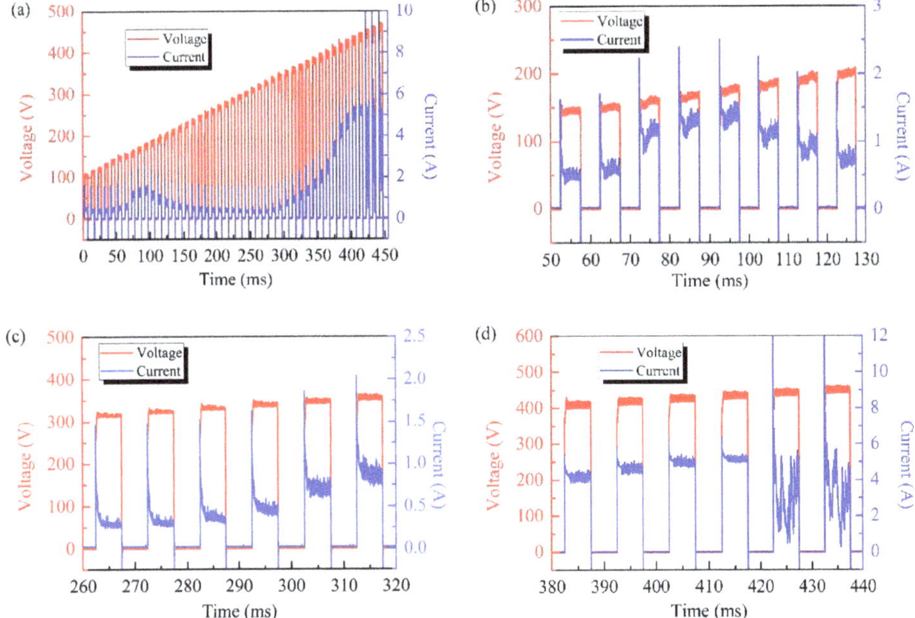

Figure 2. (**a**) Voltage and current waveforms of the initial test pulses and (**b–d**) partial waveforms.

Figure 3 depicts the waveforms of the pulses used in the test experiment and during the voltage adjustment process. When the value of the real-time feedback current was lower than half the value of I_b, a new series of test pulses (330–350 V) was applied. According to the current waveforms of the test pulses, new values of V_b (335 V) and I_b (0.6 A) were determined and they correspond to the inflection point of the current curve. At this point, a new value of the amplitude (340 V) was assigned to the voltage pulses.

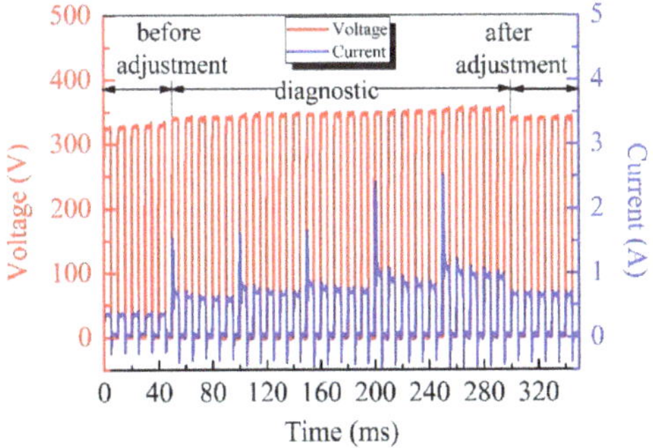

Figure 3. Voltage and current waveforms of pulses during a typical self-adaptive adjustment process.

Figure 4a shows the voltage and the current as a function of the measurement time: The voltage increased linearly in the 330–450V range and the current remained rather stable. These observations show that the current overshoot process, which occurs in the traditional voltage-control mode [6], was effectively suppressed. The current termination rule adopted in this work more effectively maintained a steady voltage in time, when compared to the more commonly used fixed-time methods. Figure 4b shows the output power curve, which was obtained experimentally: The power increased as a function of the processing time, reflecting that a higher pulse energy should be applied during the coating growth process to ensure the dielectric breakdown and a series of discharges.

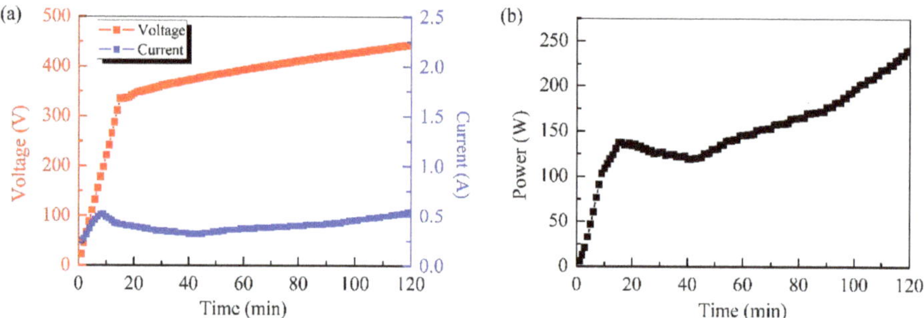

Figure 4. (**a**) Voltage–time curve, current–time curve, and (**b**) power–time curve during the self-adaptive plasma electrolytic oxidation processing of aluminum alloys.

3.2. Coating Microstructure

The surface morphology of the coating exhibits (Figure 5a) regions with elongated open pores and lighter gray areas. Open pores are generally observed for short processing times or when a low-voltage amplitude is applied to low-thickness coatings [15]. The cross-sectional image of the sample (Figure 5b) reveals that the interfaces between the substrate and the pores and between the inner and the outer layer of the coatings were characterized by a wavy profile. The estimated thickness of the coating measures 23 µm and this value was lower than the previously reported ones (100 µm) [15,24]. Figure 5c shows that the coating was mostly composed of Al, O, and Si. Moreover, the phases of the coating were mainly composed of the α-Al_2O_3 and γ-Al_2O_3 (Figure 5d), which could enhance the coating microhardness and are formed under a soft-sparking regime.

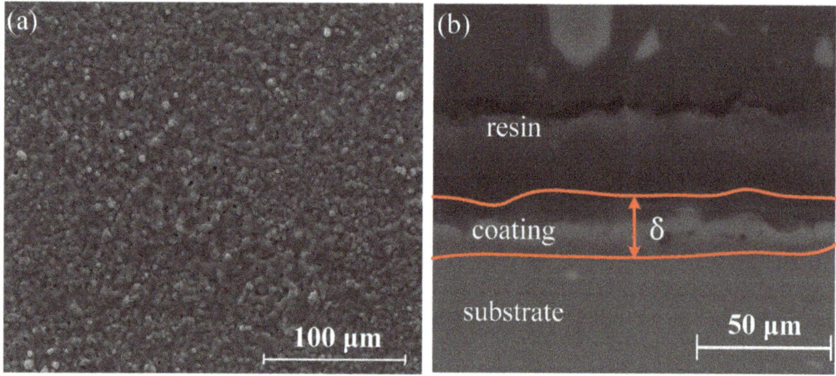

Figure 5. *Cont.*

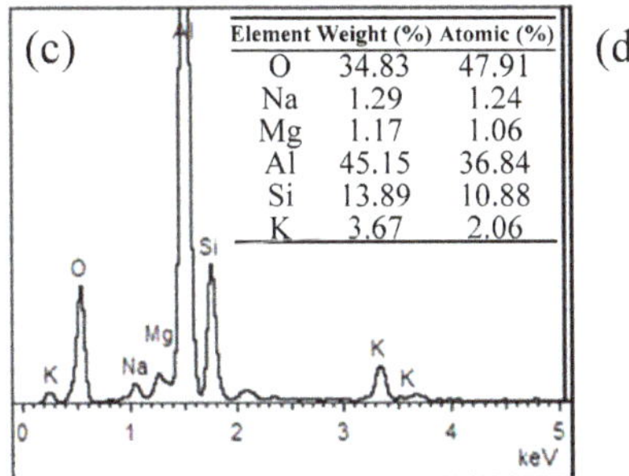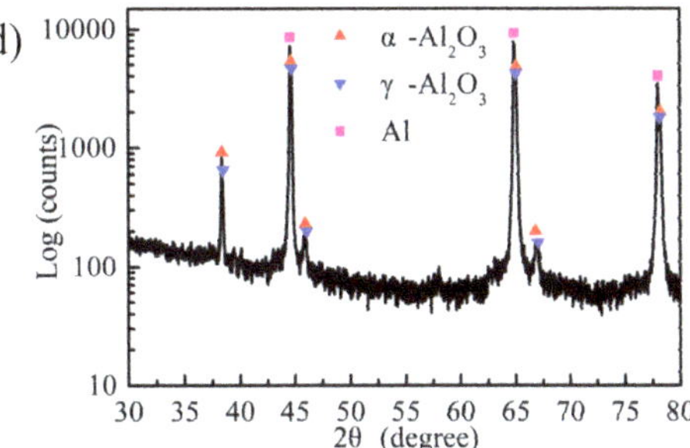

Figure 5. (**a**) Surface morphology, (**b**) cross-section, (**c**) chemical compositions, and (**d**) phases of the plasma electrolytic oxidation coatings.

3.3. Energy Consumption

The specific energy consumption (Q_c) of the plasma electrolytic oxidation process can be calculated by using Equation (4):

$$Q_c = \frac{P \cdot t}{S \cdot \delta} = \frac{1}{S \cdot \delta} \int_0^t u(t) \cdot i(t) dt, \tag{4}$$

where, P represents the power consumption, t is the total processing time, S corresponds to the superficial area of the sample, and δ is the coating thickness.

The value of Q_c estimated by using Equation (4) is 1.8 kW h m^{-2} μm^{-1}: This value is similar to that obtained in a previous study [24] (2.5–2.7 kW h m^{-2} μm^{-1}) and considerably lower than other results (26.7 kW h m^{-2} μm^{-1}) reported in the literature [28].

4. Discussion

The oxides of aluminum substrate have large band gaps. Such band gaps might be associated with an oxide structure with highly stable thermodynamic properties. It is well established that a dielectric breakdown commonly occurs across a thin oxide film located on a substrate [1]. Hence, an anodic oxidation process was performed before the plasma electrolytic oxidation processing of the sample, with the aim to promote discharge formation and enhance the dielectric breakdown strength. In order to ensure the electrons do not travel through the oxide film, the final voltage of the anodic oxidation process was set as 300 V, which is lower than the initial breakdown voltage.

The band gaps influence the electric field that tended to build up across the oxide film. A dielectric breakdown strength was normally expressed as a critical breakdown field, at which point, a discharge occurred. Hence, in order to assure enough dielectric breakdown strength, the applied voltage magnitude should be higher than the breakdown voltage. During the plasma electrolytic oxidation process, the electric fields were affected by the dielectric constant, which represents the capacity of the oxide to store electric charge. The dielectric constant increased with increasing thickness of the coating. Thus, the applied voltage should also be incremental. While the increasing voltage might show nonlinear behavior. Therefore, it is necessary to establish the relationship between the breakdown voltage and the coating thickness. The real-time pulse test technique provides a feasible approach to realize the identification of the feature information related to the coating growth.

The repeated formation of discharges on the surface of the sample is the key characteristic of a plasma electrolytic oxidation process. An individual event consisted of a complicated process such as micro-discharges, plasma channels, melting, evaporation, ejection, atomization, ionization, chemical reaction, cooling down, and overgrowth. According to previous research [1,5,10,15,20], the discharges change following a well-defined trend: The first spark, generating a discharge phenomenon, appeared by dielectric breakdown through a "weak site" in the anodic oxide film. The number of weak sites reduced with increasing thickness of the coating. With increasing voltage, the discharge color turned to orange–red, along with the emergence of more intense acoustic noise during the plasma electrolytic oxidation process. The individual discharges become less frequent but more intense when the increasing thickness of the coating, due to the reduced number of discharging sites through which the higher applied voltage needed to offer supplementary energy. These discharges have a strong tendency to occur repeatedly at particular locations—for example, they occur in 'cascades' that typically consist of hundreds of individual discharges [11]. Moreover, these discharges become more energetic and more dispersed—in terms of time and location—as the thickness increases. This is also verified by the evolving microstructure of the coating, particularly the pore content and architecture [20]. In a word, with an increasing applied voltage, the size of micropores, discharge channels, and overgrowth protrusions increased, which are mainly attributed to the different discharge energy supplied by high voltage. However, when the applied voltage is too high, negative effects on high-quality and good-performance coatings are observed. Hence, the discharge energy and coating microstructure are dependent on the electrical parameters of the process.

These results indicate that the discharge events, which occur during the self-adaptive plasma electrolytic oxidation process, mainly belong to the first three stages described in the literature [15]: Stage 1, a thin oxide film was formed and dielectric breakdown was observed; stage 2, many white sparks were evenly distributed on the entire surface of the sample; stage 3, the sparks were gradually replaced by more intense micro-discharges with yellow or orange appearance. All of the discharges mentioned above are maintained in the soft-sparking regime. When a low V_u value is chosen, the discharges occur in the soft-sparking regime and this prevents the formation of cracks and other destructive effects on the coating, but this influences the choice of the coating thickness. The solution of such a compromise lies in the use of a novel soft-sparking regime based on the bipolar pulse mode, which occurs only when the ratio between the anodic and cathodic charge is lower than one [19–21,29].

It was estimated [32] that an individual discharge energy was ~1 mJ and that the conversion rate between discharge energy and resultant volume of the coating was ~10^{13} J m^{-3}, which can also be expressed as about 3 kW h m^{-2} μm^{-1}. The specific energy consumption of the self-adaptive plasma electrolytic oxidation process was only 1.8 kW h m^{-2} μm^{-1}. Such improvement might be related to the use of the pre-anodized precursor films and the self-adaptive control of several process parameters, such as the voltage, the current, and the processing time. Recently, the precursor anodic porous films prepared by conventional anodizing were demonstrated to reduce the energy consumption and to increase the coating microhardness via promotion of the soft-sparking regime [24,25]. Meanwhile, the energy associated with each discharge may be optimally controlled via real-time precision adjustment of the process electrical parameters.

5. Conclusions

(1) A novel self-adaptive control method was used to prepare plasma electrolytic oxidation coatings on 6061 aluminum alloy samples.

(2) The feature information (V_b, I_b, V_i, and V_u) related to the coating growth was obtained from real-time feedback electrical signals generated via irregular test pulses. The process parameters (voltage, current, and processing time), which were applied during the coating preparation, were automatically adjusted according to the feature information.

(3) The coating produced via plasma electrolytic oxidation exhibited good uniformity and compactness, due to the extensive distribution of the small scattered ceramic particles. Moreover, no cracks and large pores were observed.

(4) The specific energy consumption of the self-adaptive plasma electrolytic oxidation process measured only 1.8 kW h m^{-2} μm^{-1}.

6. Patents

Cao, B.; Yang, K.; Huang Z. Adaptive control method and system for plasma electrolytic oxidation process. *China ZL* 201410036861.5, 2016.

Author Contributions: Conceptualization and Methodology, K.Y. and B.C.; Investigation, K.Y., H.H., J.C. and B.C.; Resources, K.Y., H.H. and B.C.; Writing—original draft, K.Y. and J.Z.; Writing—review & editing, J.Z., H.H. and B.C.; Funding acquisition, H.H., and J.C.

Funding: This research was funded by National Natural Science Foundation of China (grant number 51865004) and Foundation of Guizhou Educational Committee (KY [2017]106).

Acknowledgments: The authors are grateful for technical support and equipment provided by Guangzhou Jing Yuan Electrical Equipment Limited Company.

Conflicts of Interest: The authors declare no conflict of interest.

References

1. Clyne, T.W.; Troughton, S.C. A review of recent work on discharge characteristics during plasma electrolytic oxidation of various metals. *Int. Mater. Rev.* **2019**, *64*, 127–162. [CrossRef]
2. Rizwan, M.; Alias, R.; Zaidi, U.Z.; Mahmoodian, R.; Hamdi, M. Surface modification of valve metals using plasma electrolytic oxidation for antibacterial applications: A review. *J. Biomed. Mater. Res. A* **2018**, *106*, 590–605. [CrossRef] [PubMed]
3. Martin, J.; Melhem, A.; Shchedrina, I.; Duchanoy, T.; Nominé, A.; Henrion, G.; Czerwiec, T.; Belmonte, T. Effects of electrical parameters on plasma electrolytic oxidation of aluminum. *Surf. Coat. Technol.* **2013**, *221*, 70–76. [CrossRef]
4. Dehnavi, V.; Luan, B.L.; Shoesmith, D.W.; Liu, X.Y.; Rohani, S. Effect of duty cycle and applied current frequency on plasma electrolytic oxidation (PEO) coating growth behavior. *Surf. Coat. Technol.* **2013**, *226*, 100–107. [CrossRef]
5. Hussein, R.O.; Nie, X.; Northwood, D.O. An investigation of ceramic coating growth mechanisms in plasma electrolytic oxidation (PEO) processing. *Electrochim. Acta* **2013**, *112*, 111–119. [CrossRef]
6. Yang, K.; Cao, B. Electrical characteristics identification of dielectric film breakdown during plasma electrolytic oxidation process. *Mater. Lett.* **2015**, *143*, 177–180. [CrossRef]
7. Sreekanth, D.; Rameshbabu, N.; Venkateswarlu, K. Effect of various additives on morphology and corrosion behavior of ceramic coatings developed on AZ31 magnesium alloy by plasma electrolytic oxidation. *Ceram. Int.* **2012**, *38*, 4607–4615. [CrossRef]
8. Lu, X.; Mohedano, M.; Blawert, C.; Matykina, E.; Arrabal, R.; Kainer, K.U.; Zheludkevich, M.L. Plasma electrolytic oxidation coatings with particle additions—A review. *Surf. Coat. Technol.* **2016**, *307*, 1165–1182. [CrossRef]
9. Matykina, E.; Berkani, A.; Skeldon, P.; Thompson, G.E. Real-time imaging of coating growth during plasma electrolytic oxidation of titanium. *Electrochim. Acta* **2007**, *53*, 1987–1994. [CrossRef]
10. Yerokhin, A.L.; Snizhko, L.O.; Gurevina, N.L.; Leyland, A.; Pilkington, A.; Matthews, A. Spatial characteristics of discharge phenomena in plasma electrolytic oxidation of aluminum alloy. *Surf. Coat. Technol.* **2004**, *177*, 779–783. [CrossRef]
11. Yang, K.; Huang, H.S.; Chen, J.D.; Cao, B. Discharge behavior and dielectric breakdown of oxide films during single pulse anodizing of aluminum micro-electrodes. *Materials* **2019**, *12*, 2286. [CrossRef]
12. Yerokhin, A.; Mukaeva, V.R.; Parfenov, E.V.; Laugel, N.; Matthews, A. Charge transfer mechanisms underlying Contact Glow Discharge Electrolysis. *Electrochim. Acta* **2019**, *312*, 441–456. [CrossRef]
13. Troughton, S.C.; Clyne, T.W. Cathodic discharges during high frequency plasma electrolytic oxidation. *Surf. Coat. Technol.* **2018**, *352*, 591–599. [CrossRef]
14. Ko, Y.G.; Lee, E.S.; Shin, D.H. Influence of voltage waveform on anodic film of AZ91 Mg alloy via plasma electrolytic oxidation: Microstructural characteristics and electrochemical responses. *J. Alloy. Compd.* **2014**, *586*, S357–S361. [CrossRef]
15. Dehnavi, V.; Luan, B.L.; Liu, X.Y.; Shoesmith, D.W.; Rohani, S. Correlation between plasma electrolytic oxidation treatment stages and coating microstructure on aluminum under unipolar pulsed DC mode. *Surf. Coat. Technol.* **2015**, *269*, 91–99. [CrossRef]
16. Hussein, R.O.; Northwood, D.O.; Nie, X. The influence of pulse timing and current mode on the microstructure and corrosion behavior of a plasma electrolytic oxidation (PEO) coated AM60B magnesium alloy. *J. Alloy. Compd.* **2012**, *541*, 41–48. [CrossRef]

17. Hussein, R.O.; Northwood, D.O.; Su, J.F.; Nie, X. A study of the interactive effects of hybrid current modes on the tribological properties of a PEO (plasma electrolytic oxidation) coated AM60B Mg-alloy. *Surf. Coat. Technol.* **2013**, *215*, 421–430. [CrossRef]

18. Asri, R.I.M.; Harun, W.S.W.; Samykano, M.; Lah, N.A.C.; Ghani, S.A.C.; Tarlochan, F.; Raza, M.R. Corrosion and surface modification on biocompatible metals: A review. *Mat. Sci. Eng. C* **2017**, *77*, 1261–1274. [CrossRef]

19. Rogov, A.B.; Yerokhin, A.; Matthews, A. The role of cathodic current in plasma electrolytic oxidation of aluminum: Current density scanning waves on complex-shape substrates. *J. Phys. D Appl. Phys.* **2018**, *51*, 405303. [CrossRef]

20. Zou, Y.C.; Wang, Y.M.; Sun, Z.D.; Cui, Y.; Jin, T.; Wei, D.Q.; Ouyang, J.H.; Jia, D.C.; Zhou, Y. Plasma electrolytic oxidation induced 'local over-growth' characteristic across substrate/coating interface: Effects and tailoring strategy of individual pulse energy. *Surf. Coat. Tech.* **2018**, *342*, 198–208. [CrossRef]

21. Hakimizad, A.; Raeissi, K.; Santamaria, M.; Asghari, M. Effects of pulse current mode on plasma electrolytic oxidation of 7075 Al in Na_2WO_4 containing solution: From unipolar to soft-sparking regime. *Electrochim. Acta* **2018**, *284*, 618–629. [CrossRef]

22. Tsai, D.S.; Chen, G.W.; Chou, C.C. Probe the micro arc softening phenomenon with pulse transient analysis in plasma electrolytic oxidation. *Surf. Coat. Technol.* **2019**, *357*, 235–243. [CrossRef]

23. Rogov, A.B.; Yerokhin, A.; Matthews, A. The role of cathodic current in plasma electrolytic oxidation of aluminum: Phenomenological concepts of the "soft-sparking" mode. *Langmuir* **2017**, *33*, 11059–11069. [CrossRef] [PubMed]

24. Matykina, E.; Arrabal, R.; Pardo, A.; Mohedano, M.; Mingo, B.; Rodríguez, I.; González, J. Energy-efficient PEO process of aluminum alloys. *Mater. Lett.* **2014**, *127*, 13–16. [CrossRef]

25. Matykina, E.; Arrabal, R.; Mohedano, M.; Mingo, B.; Gonzalez, J.; Pardo, A.; Merino, M.C. Recent advances in energy efficient PEO processing of aluminum alloys. *Trans. Nonferr. Metals Soc. China* **2017**, *27*, 1439–1454. [CrossRef]

26. Troughton, S.C.; Nominé, A.; Dean, J.; Clyne, T.W. Effect of individual discharge cascades on the microstructure of plasma electrolytic oxidation coatings. *Appl. Sur. Sci.* **2016**, *389*, 260–269. [CrossRef]

27. Rokosz, K.; Hryniewicz, T.; Gaiaschi, S.; Chapon, P.; Raaen, S.; Matýsek, D.; Dudek, L.; Pietrzak, K. Novel porous phosphorus–calcium–magnesium coatings on titanium with copper or zinc obtained by DC plasma electrolytic oxidation fabrication and characterization. *Materials* **2018**, *11*, 1680. [CrossRef]

28. Yerokhin, A.L.; Shatrov, A.; Samsonov, V.; Shashkov, P.; Pilkington, A.; Leyland, A.; Matthews, A. Oxide ceramic coatings on aluminum alloys produced by a pulsed bipolar plasma electrolytic oxidation process. *Surf. Coat. Technol.* **2005**, *199*, 150–157. [CrossRef]

29. Mécuson, F.J.; Czerwiec, T.; Henrion, G.; Belmonte, T.; Dujardin, L.; Viola, A.; Beauvir, J. Tailored aluminum oxide layers by bipolar current adjustment in the plasma electrolytic oxidation (PEO) process. *Surf. Coat. Technol.* **2007**, *201*, 8677–8682. [CrossRef]

30. Du, K.Q.; Guo, X.H.; Guo, Q.Z.; Wang, F.H.; Tian, Y. A monolayer PEO coating on 2024 Al alloy by transient self-feedback control mode. *Mater. Lett.* **2013**, *91*, 45–49. [CrossRef]

31. Parfenov, E.V.; Yerokhin, A.; Nevyantseva, R.R.; Gorbatkov, M.V.; Liang, C.J.; Matthews, A. Towards smart electrolytic plasma technologies: An overview of methodological approaches to process modelling. *Surf. Coat. Technol.* **2015**, *269*, 2–22. [CrossRef]

32. Troughton, S.C.; Nominé, A.; Nominé, A.V.; Henrion, G.; Clyne, T.W. Synchronized electrical monitoring and high-speed video of bubble growth associated with individual discharges during plasma electrolytic oxidation. *Appl. Sur. Sci.* **2015**, *359*, 405–411. [CrossRef]

Article

Discharge Behavior and Dielectric Breakdown of Oxide Films during Single Pulse Anodizing of Aluminum Micro-Electrodes

Kai Yang [1,2], Haisong Huang [1], Jiadui Chen [1] and Biao Cao [2,*]

[1]   Key Laboratory of Advanced Manufacturing Technology of the Ministry of Education, Guizhou University, Guiyang 550025, China
[2]   College of Mechanical and Automotive Engineering, South China University of Technology, Guangzhou 510640, China
*   Correspondence: mebcao@scut.edu.cn; Tel.: +86-138-2504-3678

Received: 24 June 2019; Accepted: 15 July 2019; Published: 17 July 2019

Abstract: Micro-arc discharge events and dielectric breakdown of oxide films play an important role in the formation process of plasma electrolytic oxidation coating. Single pulse anodization of micro-electrodes was employed to study the discharge behavior and dielectric breakdown of oxide films deposited on aluminum in an alkaline silicate electrolyte. Voltage and current waveforms of applied pulses were measured and surface morphology of micro-electrodes was characterized from images obtained using scanning electron microscope (SEM). A feasible identification method for the critical breakdown voltage of oxide film was introduced. Different current transients of voltage pulses were obtained, depending on applied pulse voltage and duration. In addition, the active capacitive effect and complex non-linear nature of plasma electrolytic oxidation process is confirmed using dynamic electrical characteristic curves. A good correlation between the pulse parameters and shape of discharge channels was observed. Circular opened pores were found to close with increasing potential and pulse width. Finally, the characteristic parameters of a single discharge event were estimated.

Keywords: plasma electrolytic oxidation; electrical characteristic; anodizing; SEM; aluminum

---

1. Introduction

Plasma electrolytic oxidation (PEO), often also known as micro-arc oxidation (MAO), is a novel environmentally friendly surface modification technique used for the production of ceramic coatings on a variety of light metals, such as Al, Mg, Ti and their alloys [1,2]. PEO is a complex and highly non-linear process due to the electrical, thermal, and plasma-chemical reactions in the electrolyte. Over the last decade, some prominent researches have been reported in literature related to the formation mechanism of PEO coatings. Various approaches have been reported in literature to identify the mechanism for the formation of the coating due to PEO, such as thickness measurement [3], element tracer [4,5], two-step oxidation method [6], and optical emission spectroscopy (OES) [7]. Nevertheless, it is still difficult to explain such a complex physical and chemical process during plasma discharge and significant work still needs to be undertaken to explain the formation mechanism of PEO coatings.

During PEO, dielectric breakdown of the growing oxide coatings at high voltage electrolysis results in a large number of short-lived micro-discharges [8]. Those discharge events mean not a single discharge but a cascade of smaller discharges that are bundled together in space and time. These distinct discharge events play an important role in the formation mechanism of the coating and strongly affect the microstructure and properties of the coating [9]. Due to the extreme non-linearity associated with the plasma discharge, monitoring the evolution of an individual discharge event would aid in understanding the coating mechanism and allow optimization of the production process.

Recently, many methods, such as optical [10–12], spectral [12–16], electrical [16,17], frequency response [18,19], and acoustic [20], have been used to measure dynamic parameters of plasma discharge during PEO. Reported characteristic parameters, including duration, current level, apparent radii, event rate, temperature, and electron density, about PEO discharge events are summarized [10–17,21] as follows: Typical discharges are now known to occur in prolonged sequences ('cascades') at particular locations, and to have lifetimes of the order of a few tens to a few hundreds of microseconds, with 'incubation' periods between them of around a few hundred μs to a ms or two. Discharge currents are typically several tens of mA, discharge energies a few mJ and diameters of core discharge channels a few tens of μm. The frequency of plasma discharges per unit time and area was determined throughout processing, falling from initial values in the range 300–350 mm^{-2} s^{-1} to fewer than 50 mm^{-2} s^{-1} after 1000 s of processing. Plasma temperatures have been estimated via optical spectroscopy to range from about 4000 to 12,000 K, with some indications of a higher temperature core and a lower temperature surrounding region. Corresponding electron densities typically range from ~10^{15} to 10^{18} cm^{-3}.

It can be seen that these monitoring techniques are effective in distinguishing certain basic characteristics of PEO discharge events. However, values of these characteristic parameters are spread across a wide range, sometimes differing by several orders of magnitude. For example, the lifetimes of discharges revealed by fast video imaging were in the range of 0.05–4 ms during AC PEO of magnesium alloys [10] and evaluations based on the spectroscopic method for the Al 1100 alloy show electron temperatures to be in the range of 3500–9000 K for a unipolar current mode and in the range of 3500–6000 K for a bipolar current mode [14]. This is probably the result of a low-resolution tool or inappropriate measurement procedure and data processing algorithm.

Special experimental methods, such as small area in-situ testing [16] and single pulse anodizing [22–24], have been designed to study discharge events and dielectric breakdown during PEO. The present work builds upon single pulse anodizing of aluminum micro-electrodes. Here, a partial aim is to produce a clear set of conclusions related to the discharge behaviors. In addition, an improved understanding of the effect of plasma discharge on subsequent oxide film formation is sought. Electrical characteristic monitoring and scanning electron microscopy (SEM) were used to study the effect of single pulse anodizing on aluminum micro-electrodes along with the effects of pulse parameters on surface morphology. Parameters such as current level, duration, diameter, and spatial distribution ratio were estimated from the current waveforms.

2. Materials and Methods

2.1. Materials and Pre-Treatment

The experimental platform used to study PEO is shown in Figure 1. A home-made 20 kW pulse power supply was used, which could provide a DC voltage ranging from 0 to 1000 V and pulses with the same magnitude and frequencies up to 20 kHz [25]. In this study, small area sample monitoring of currents in individual discharges was used to study the nature of these discharges and their distributions in time and position. In order to provide a sufficiently small area, micro-electrodes made of aluminum wires were chose as specimens. Pure aluminum wires (99.99%) with 0.6 mm diameter were anodized in electrolyte, which was composed of commercial distilled water, 5 g/L potassium hydroxide (KOH) and 10 g/L sodium silicate (Na_2SiO_3). Experiments were carried out at 25 °C in a 2-liter water-cooled stainless steel tank, which also served as the counter-electrode.

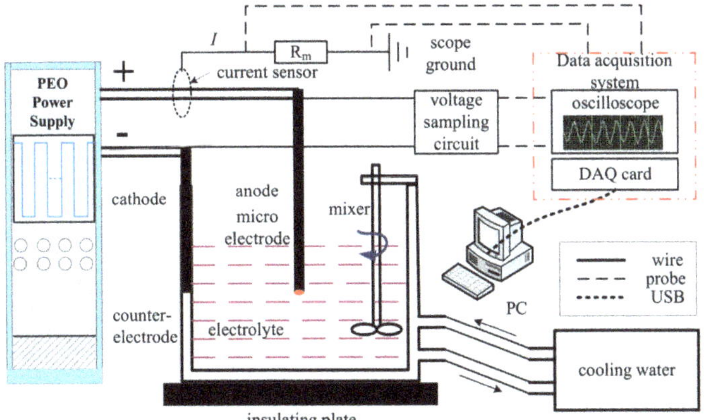

Figure 1. Schematic of the experimental platform for plasma electrolytic oxidation (PEO).

Aluminum wires were first anodized up to DC 550 V at a rate of 8 V/s, to cover the entire metal surface with a thick anodic oxide film. Next, these wires were embedded in an epoxy resin. To expose a clean and relatively flat surface, the embedded wires were sectioned with a slicing knife. All specimens were rinsed under running water to keep the tip surface clean before the next process.

2.2. Single Pulse Anodizing

In our experiments, every single pulse anodizing process was carried out with an individual specimen and the interval time between adjacent experiments was set for 5 min. All single pulse anodizing processes were carried out in two steps. Initially, pre-deposited oxide films on the tip surface were obtained by anodizing with a DC voltage sweep up to 300 V at a rate of 8 V/s. Then, a single pulse voltage with amplitude in the range of 325–525 V and a pulse width of 100–5000 μs was applied to cause dielectric breakdown.

In addition, a longer voltage pulse with a magnitude at 500 V and pulse duration of 50 ms was applied on these pre-deposited oxide films to estimate electrical characteristic parameters, such as current level, duration, diameter, and spatial distribution ratio.

2.3. Data Monitoring and Post-Processing

For pre-treatment and single pulse anodizing process, voltage and current data were recorded at a sampling rate of 2 MHz using a data acquisition card (National Instruments, PCI-6133, Austin, TX, USA) controlled by the LabVIEW software installed in PC (Thinkpad T580). Current signal was sampled by a hall current sensor (CSM002A) and voltage signal between anode and cathode electrodes was detected by probe of oscilloscope (DPO 3014) directly. Meanwhile, real-time waveforms of current and voltage were displayed on the screen of an oscilloscope. Surfaces of the specimens were observed using a LEO1530 VP scanning electron microscope.

Raw voltage and current data had considerable electrical noise due to mechanical vibration, electron avalanche, dielectric breakdown and high frequency electronic switching. As a result, it was necessary to carry out post-processing for data. The data were first filtered with a low pass filter with a cut-off frequency of 50 kHz. This allowed high frequency noise due to the electronic switching to be eliminated. Next, this data was smoothened using moving average filter. Finally, an appropriate scale function and wavelet basis function were adopted to extract random noise signal.

3. Results

3.1. Identification of Critical Breakdown Voltage

Figure 2a shows the partial voltage and current waveforms during the pre-treatment process of aluminum wires. In addition, dielectric breakdown of the oxide films when the potential was increased to approximately 350 V is shown in the figure. Within t_r, a rise of potential results in an abrupt increase in current due to the active capacitive effect of the electrolyte load. After reaching a peak, the current decreases exponentially to a relatively stable level (within t_s). When potential is lower than 340 V, stable levels are almost equivalent to 0.2 A. Above 350 V, the stable level increases with potential with an oscillating current waveform. This probably indicates that dielectric breakdown of oxide films occurred. Before dielectric breakdown, the applied voltage is not high enough to cause discharge and the load keeps in a stable high impedance. So, the current waveform is relatively smooth.

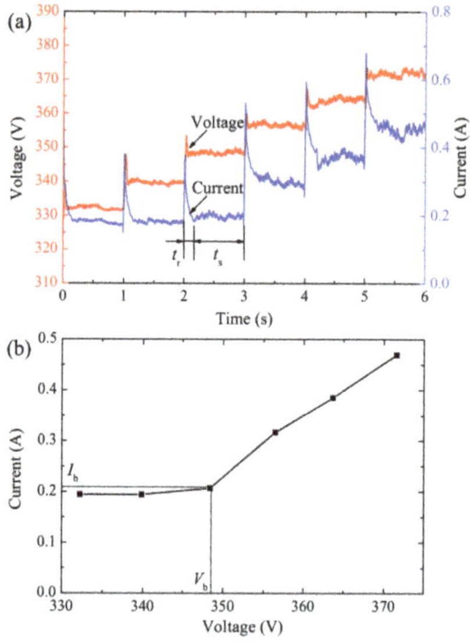

Figure 2. (**a**) Partial voltage and current waveforms during pre-treatment process with voltage sweep at 8 V/s and (**b**) their voltage-current characteristic curve.

To better understand this phenomenon, appropriate RMS value of the current and voltage approximations were calculated for each step. Stable value of current (within t_s) was chosen to improve approximation accuracy for the voltage-current characteristics shown in Figure 2b.

3.2. Electrical Characteristics of Single Pulse Anodizing

To further understand the influence of pulse voltage on electrical characteristics of single pulse anodizing of aluminum micro-electrode, a group of single pulses with different voltages were applied for 100 μs. Figure 3a depicts the current transients due to these pulses. A peak current (I_p) was observed within t_p and increases with the rise of the pulse voltage. Next, a sharp decrease is seen for the current and it reaches a stable and nearly constant current (I_c) for different level, from 0.2 to 1.2 A, flows within t_c. Peak current due to capacitive load effect produces a large reactive component of the current. Magnitude of I_p and current transition time t_p were related to pulse potential. A higher potential resulted in a larger I_p and longer t_p. An increase in the value of I_c was also observed with an increase in the applied pulse voltage. However, transient behavior was different at 375 V. Current oscillations with a large amplitude were observed and could be attributed to the tiny flashing sparks.

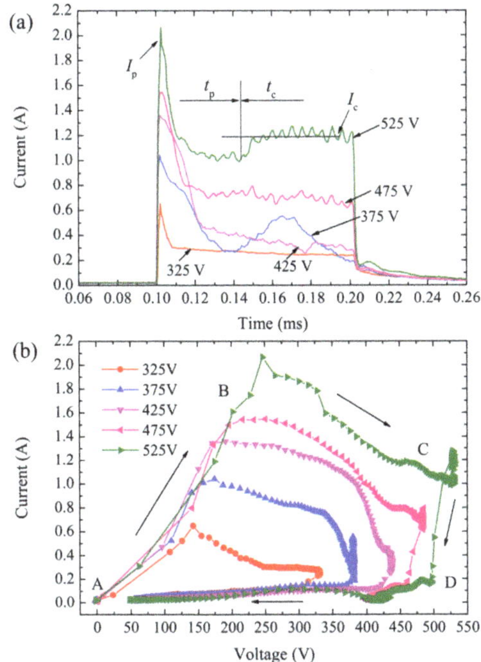

Figure 3. Electrical characteristic curves of single pulse anodizing under different potentials, (**a**) current transients and (**b**) *V–I* curve.

Figure 3b shows the dynamic *V–I* characteristic curves of single pulses, which is characterized by a loop from A to D along the marked direction. An abrupt current increase occurred due to a rapid initial voltage rise (path A-B), which was followed by a decrease in current to a lower level until the voltage reached a target value (path B-C). Next, a gradual and subtle change in voltage and current values were observed (path C-D). Finally, a sudden drop in the current decreases to zero was observed with voltage decreasing at a lower rate (path D-A). Shape and scope of these curves were affected by magnitude and change rate of the pulse potential. Higher potential resulted in longer loop line and larger loop area.

Similar current transients were observed for pulse voltages ranging from 325 to 525 V with a pulse duration of 500 μs (shown in Figure 4). With an increase in the pulse duration, differences among these pulses were prominent. For the pulse potential lower than 375 V, current transient was relatively smooth and decreased almost linearly with time. When the pulse potential was higher than 375 V, oscillation due to spark discharge was observed in the current transient due to dielectric breakdown of the coatings. For the same pulse potential, the effective value of the pulse current is inversely proportional to the pulse duration. This suggests the existence of a more reactive component in the overall current corresponding to the shorter pulses, due to the initial capacitor charging process.

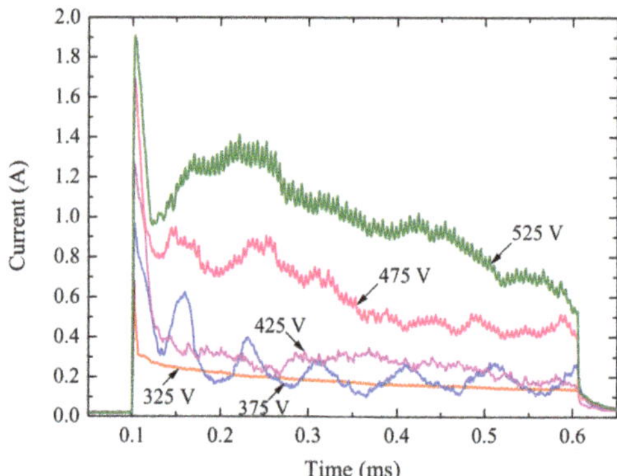

Figure 4. Current transient curves for single pulse anodizing under different potentials.

3.3. Effects of Pulse Parameters on Surface Morphology

Surface morphology after single pulse anodizing varies for different pulse voltages. This reflects voltage-dependent discharge characteristics. Figure 5a shows that the surface of the micro-electrode before single pulse anodizing was relatively flat and only several knife marks were visible at a higher resolution. Surfaces of micro-electrodes after single pulse anodizing are shown in Figure 5b–d. Figure 5b shows the surface after anodization using 375 V/100 µs pulse. Similar surfaces were observed in Figure 5c,d, which show anodization surfaces for 425 V/100 µs and 475 V/100 µs, respectively.

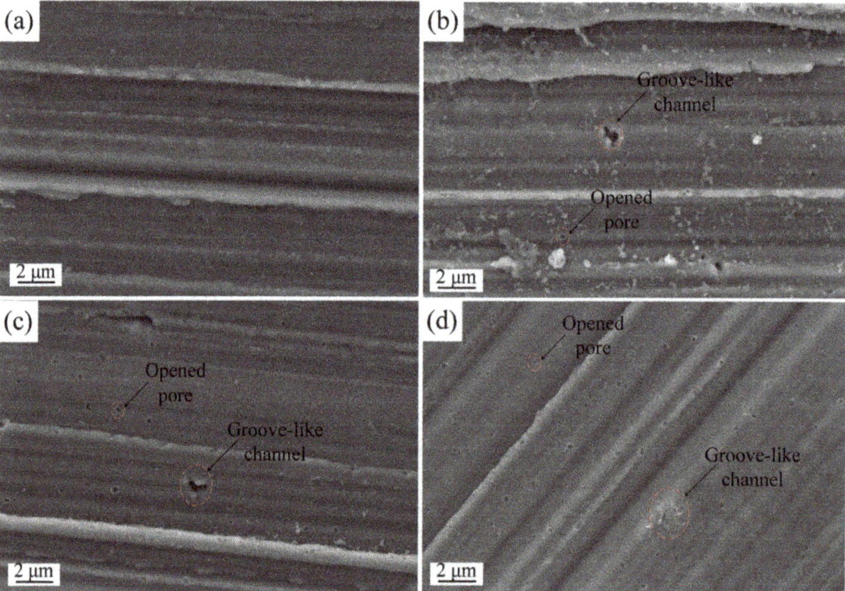

Figure 5. Scanning electron microscope (SEM) images of micro-electrode surfaces, (**a**) before single pulse anodizing, (**b**) anodizing with 375 V/100 µs pulse, (**c**) anodizing with 425 V/100 µs pulse, (**d**) anodizing with 475 V/100 µs pulse.

Figure 6 shows the surface SEM images of the specimens after applying a 400 V pulse for different pulse widths. When shorter pulses of 100 µs and 500 µs were applied, circular discharge channels with opened pores were found in addition to the groove-like channels, as shown in Figure 6a,b. These surface morphologies are similar to those shown in Figure 5. However, after breakdown, surface morphology for longer duration pulses was found to be significantly different, as shown in Figure 6c,d.

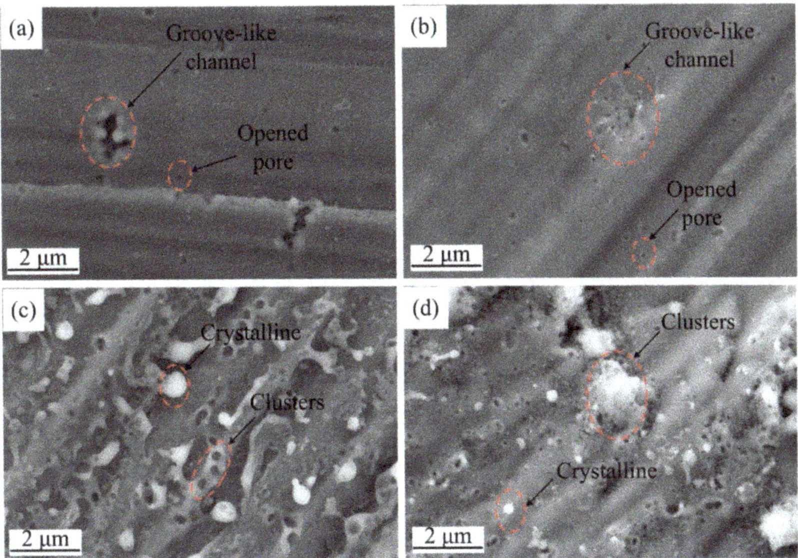

Figure 6. SEM images of micro-electrodes obtained in different conditions, (**a**) anodizing with 400 V/100 µs pulse, (**b**) anodizing with 400 V/500 µs pulse, (**c**) anodizing with 400 V/2 ms pulse, (**d**) anodizing with 400 V/5 ms pulse.

3.4. Characteristic Parameters of Discharge Event

Figure 7a shows the raw voltage and current waveforms during anodizing with 500 V/50 ms pulse. Multiple peaks and low baseline were observed in the current waveform. Each peak in the current was accompanied by a decrease in voltage, which recovered during the periods of low current, until the next series of peaks in the current. A magnified section of the current transient is shown in Figure 7b. The discrete nature of individual peaks in the current is evident here. In addition, some cascades appear as the superposition of two current peaks and these probably represent the occurrence of a second discharge event with the first one still ongoing.

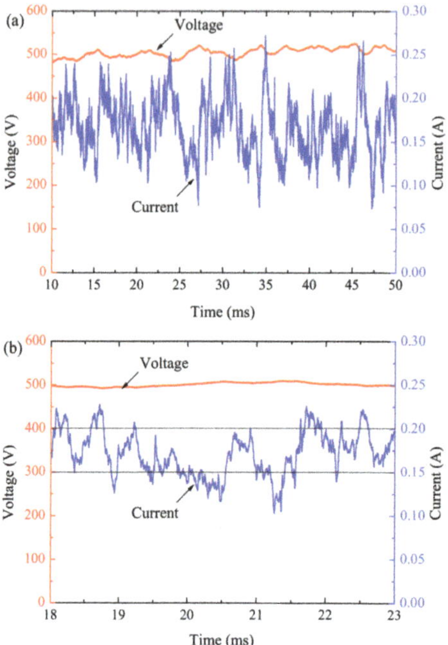

Figure 7. Voltage and current waveforms acquired during anodizing with 500 V/50 ms pulse, at (**a**) low resolution (period of 40 ms) and (**b**) high resolution (period of 5 ms).

4. Discussion

An inflection point can be clearly seen in the voltage-current characteristic curve, as shown in Figure 2b. According to our previous work [26], the voltage value of the inflection point can be considered as the critical breakdown voltage (V_b). Similarly, the corresponding current is the critical breakdown current (I_b). Increase in the current values with increasing voltage suggests breakdown of the anodic oxides occurs continuously for the whole duration. Although breakdown voltage is dependent on layer thickness and some other characteristics, this voltage-current characteristic curve provides a feasible method to identify breakdown voltage of growing oxide film during PEO.

There is a noticeable peak current at the beginning of every single pulse, which was not associated with discharges. This phenomenon appears to be consistent with a hypothesis of active–capacitive load behavior for the PEO process, which was studied systematically in literature [18]. The simple equivalent circuit of PEO load was composed of series-parallel capacitor and resistance, which can have a physical interpretation as the equivalent capacitance of the oxide film and the equivalent resistance introduced by the current passing through the discharges, respectively. The sharp decrease of current was due to the sudden increase of load impedance after the capacitor charging process. In Figures 3a and 4, current oscillations observed at 375 V reflected intense changes of load impedance, caused by dielectric breakdown of oxide film along with abundant randomly flashing sparks. In Section 3.1 and our previous work [26], the initial critical breakdown voltage of oxide film was about 350 V. So, we speculate that the voltage region for these oscillations is among 350–400 V.

It is assumed that electrons cannot flow through the oxide, so the field across it builds up as the applied voltage is raised, and it may reach the breakdown level for the oxide, at which point a discharge occurs [21]. So, the key issue is whether the electric field across the residual oxide film reaches the level necessary for dielectric breakdown. During PEO, the breakdown voltage of oxide film increases with growing layer thickness. In other hand, PEO process can be divided into several stages according to the evolution of the discharges [27]. At the early stage of the process (step 1), the growing oxide film breaks down due to an increase in the applied voltage. This stage is characterized by sparks flashing randomly all over the aluminum alloy surface. Following this stage, sparks progressively change to micro-arcs (step 2). Finally, after the end of the rapid micro-arcs decrease, step 3 is characterized by a few strong remaining arcs that appear on the surface. The average lifetime of the discharges increased with increased voltage and processing time in the main period of PEO coating growth [8]. So, we try to establish the correlation between electrical characteristics and physical phenomena, including the oxide film and discharges. In the single pulse anodizing process, all voltage pulses were applied to the same thin initial oxide film. Pulse energies, which depend on the voltage, current and pulse duration, impact the dielectric breakdown of oxide film and discharge shape significantly. The higher pulse energy results in stronger dielectric breakdown strength and more powerful discharge behavior.

A comparison of the surface morphology in Figure 5a with Figure 5b–d reveals some distinct changes on the surface of micro-electrodes. Dielectric breakdown due to the single voltage pulse produced a number of isolated discharge channels on this anodic oxide film. Circular opened discharge channels, tens to a few hundreds of nm in size, were found to be surrounded by a band of evidently thicker film material. In addition, large surface regions without discharge pores were also observed. The quantity and size of opened discharge pores increased with an increase in potential. This occurs because with higher electric field strength applied to the oxide film, a larger driving force is available for tunneling ionization. Furthermore, some groove-like discharge channels, which consist of numerous opened and closed discharge pores, were found on the surface, as shown in Figure 5d. Both pores and discharge channels were randomly distributed on the metal surface. This implies the irregular formation of barrier-type anodic oxide films. When numerous narrow pores overlapped in a small region, a groove-like discharge channel was formed. In some case, a few discharge pores were found to be coated with oxide produced by the repeating plasma discharge events during pulse anodizing, resulting in closed pores.

The surface morphology was also found to be dependent on pulse width to some extent. As shown in Figure 6c,d, it can be seen that the entire surface was covered with a thicker layer of anodic oxide film, modified by dielectric breakdown. The oxide surface was rough and a number of bright spots with sizes of ~1 µm or less were observed. Electric field strength between the electrolyte and metal increased with time due to continuous electron transfer. This higher stored energy makes it easier to breakdown the film. Moreover, temperature increase in the groove-like discharge channel regions could result in transforming amorphous alumina to crystalline, which usually has a higher chemical stability [22]. The closure of opened discharge pores appears to be more obvious with increasing pulse width in comparison to pulse magnitude. For both 2 and 5 ms pulse widths, numerous discharge channels were found on the surface. The size of these discharge clusters (circled in Figure 6c,d) on the oxide surface is proportional to the pulse width and an increased size was observed for longer pulse widths.

Some characteristic parameters can be obtained from Figure 7. However, difficulty arises with the identification of the start and end of each discharge event. Using the method mentioned in literature [16], we hypothesized that a peak current (I_p) and a bottom current (I_b) represent a start and end of discharge event respectively. Two current thresholds were set to estimate the total number of discharge events, 0.2 and 0.15 A. When the current exceed 0.2 A, it can be considered as a peak current of a discharge event. When the current value is lower than 0.15 A, it can be considered as the baseline current of a discharge event. Discharges were considered to initiate when the current reaches the peak and terminates when it reaches the baseline current.

Current level (I_{event}), duration (t_d), diameter (φ), and spatial distribution ratio (D_n) of discharge can be estimated using Equations (1)–(6):

$$N_{total} = \frac{1}{2}\left(N_p + N_b\right) \tag{1}$$

$$I_{\text{event}} = \sum_{n=0}^{N_{\text{total}}} \frac{1}{N_{\text{total}}} \left(I_{\text{p}} - I_{\text{b}} \right) \tag{2}$$

$$t_{\text{gap}}^{\text{n}} = \left| t_{\text{p}}^{\text{n}} - t_{\text{b}}^{\text{n}} \right| \tag{3}$$

$$t_{\text{d}} = \sum_{n=0}^{N_{\text{total}}} \frac{1}{N_{\text{total}}} t_{\text{gap}}^{\text{n}} \tag{4}$$

$$D_{\text{n}} = N_{\text{total}} / \left(S_{\text{area}} \cdot t_{\text{pulse}} \right) \tag{5}$$

$$\varphi = 2 \sqrt{\frac{S_{\text{area}}}{(D_{\text{n}} \cdot \pi)}} \tag{6}$$

In Equations (1)–(6): N_{p}, N_{b} and N_{total} represent the number of I_{p}, I_{b} and total discharge events respectively; t_{gap} represents the time interval between adjacent peak time (t_{p}) and bottom time (t_{b}); S_{area} represents the superficial area of sample; t_{pulse} represents the pulse duration.

To eliminate errors due to experimental deviation, a sample size of ten for the same pulse anodizing was carried out. All the characteristic parameters (shown in Table 1) were obtained by calculating the average value of the results of these experiments.

Table 1. Estimated characteristic parameters of individual discharge events.

Current Level (I_{event}, mA)	Duration (t_{d}, ms)	Spatial Distribution Ratio (D_{n}, mm$^{-2}\cdot$ms^{-1})	Diameter (φ, μm)
97.1	0.94	4.69	369.4

Although the characteristic parameters listed in Table 1 are similar to those results reported in section '1. Introduction', experimental methods and monitoring techniques need to be improved to acquire more accurate data. In addition, a larger sample size of experiments should be carried out to confirm these results.

5. Conclusions

(1) Characteristic phenomenon, with the occurrence of an abrupt current increase when potential reaches a critical value, shown in the voltage-current characteristic curve, allows a feasible identification method for critical breakdown voltage of oxide film.

(2) Magnitude of peak current (I_{p}), current transition time (t_{p}), and constant current (I_{c}) of the single pulse anodizing process were determined by pulse potential. Dynamic V-I characteristic curves for single pulses were characterized by loop circle with the length and area of this loop circle being related to pulse potential. With the increase of pulse width, the effective value of the current becomes smaller.

(3) Characteristic shapes of discharge channels were observed under single pulse anodizing and were correlated with the pulse parameters. Isolated circular opened pores were primarily found under shorter and lower voltage pulses. In contrast, groove-like discharge channels formed in addition to opened pores under longer and higher voltage pulses. Opened discharge pores were found to close within the groove-like discharge channels region with increasing pulse width.

(4) Approximate characteristic parameters of individual discharge events were estimated for individual discharge events of short duration (~hundreds of μs), low current (~tens of mA), and small size (~hundreds of μm).

6. Patents

Cao, B.; Yang, K.; Huang Z. Adaptive control method and system for plasma electrolytic oxidation process. *China ZL* 201410036861.5, 2016.

Author Contributions: Conceptualization and Methodology, K.Y. and B.C.; Investigation, K.Y., H.H., J.C. and B.C.; Resources, K.Y., H.H. and B.C.; Writing—original draft, K.Y.; Writing—review & editing, H.H. and B.C.; Funding acquisition, H.H., and J.C.

Funding: This research was funded by National Natural Science Foundation of China (grant number 51865004) and Foundation of Guizhou Educational Committee (KY [2017]106).

Acknowledgments: The authors are grateful for technical support and equipment provided by Guangzhou Jing Yuan Electrical Equipment Limited Company.

Conflicts of Interest: The authors declare no conflict of interest.

References

1. Parfenov, E.V.; Yerokhin, A.; Nevyantseva, R.R.; Gorbatkov, M.V.; Liang, C.J.; Matthews, A. Towards smart electrolytic plasma technologies: An overview of methodological approaches to process modelling. *Surf. Coat. Technol.* **2015**, *269*, 2–22. [CrossRef]

2. Rokosz, K.; Hryniewicz, T.; Gaiaschi, S.; Chapon, P.; Raaen, S.; Matýsek, D.; Dudek, L.; Pietrzak, K. Novel Porous Phosphorus–Calcium–Magnesium Coatings on Titanium with Copper or Zinc Obtained by DC Plasma Electrolytic Oxidation Fabrication and Characterization. *Materials* **2018**, *11*, 1680. [CrossRef] [PubMed]

3. Hussein, R.O.; Nie, X.; Northwood, D.O. An investigation of ceramic coating growth mechanisms in plasma electrolytic oxidation (PEO) processing. *Electrochim. Acta* **2013**, *112*, 111–119. [CrossRef]

4. Matykina, E.; Arrabal, R.; Scurr, D.J.; Baron, A.; Skeldon, P.; Thompson, G.E. Investigation of the mechanism of plasma electrolytic oxidation of aluminium using ^{18}O tracer. *Corros. Sci.* **2010**, *52*, 1070–1076. [CrossRef]

5. Liu, X.; Li, G.; Xia, Y. Investigation of the discharge mechanism of plasma electrolytic oxidation using Ti tracer. *Surf. Coat. Technol.* **2012**, *206*, 4462–4465. [CrossRef]

6. Liu, X.; Zhu, L.; Liu, H.; Li, W. Investigation of MAO coating growth mechanism on aluminum alloy by two-step oxidation method. *Appl. Surf. Sci.* **2014**, *293*, 12–17. [CrossRef]

7. Wang, L.; Chen, L.; Yan, Z.; Fu, W. Optical emission spectroscopy studies of discharge mechanism and plasma characteristics during plasma electrolytic oxidation of magnesium in different electrolytes. *Surf. Coat. Technol.* **2010**, *205*, 1651–1658. [CrossRef]

8. Matykina, E.; Berkani, A.; Skeldon, P.; Thompson, G.E. Real-time imaging of coating growth during plasma electrolytic oxidation of titanium. *Electrochim. Acta* **2007**, *53*, 1987–1994. [CrossRef]

9. Cheng, Y.; Wu, F.; Matykina, E.; Skeldon, P.; Thompson, G.E. The influences of micro discharge types and silicate on the morphologies and phase compositions of plasma electrolytic oxidation coatings on Zircaloy-2. *Corros. Sci.* **2012**, *59*, 307–315. [CrossRef]

10. Arrabal, R.; Matykina, E.; Hashimoto, T.; Skeldon, P.; Thompson, G.E. Characterization of AC PEO coatings on magnesium alloys. *Surf. Coat. Technol.* **2009**, *203*, 2207–2220. [CrossRef]

11. Yerokhin, A.L.; Snizhko, L.O.; Gurevina, N.L.; Leyland, A.; Pilkington, A.; Matthews, A. Spatial characteristics of discharge phenomena in plasma electrolytic oxidation of aluminum alloy. *Surf. Coat. Technol.* **2004**, *177–178*, 779–783. [CrossRef]

12. Stojadinović, S.; Jovović, J.; Petković, M.; Vasilić, R.; Konjević, N. Spectroscopic and real-time imaging investigation of tantalum plasma electrolytic oxidation (PEO). *Surf. Coat. Technol.* **2011**, *205*, 5406–5413. [CrossRef]

13. Me´cuson, F.; Czerwiec, T.; Belmonte, T.; Dujardin, L.; Viola, A.; Henrion, G. Diagnostics of an electrolytic microarc process for aluminum alloy oxidation. *Surf. Coat. Technol.* **2005**, *200*, 804–808. [CrossRef]

14. Hussein, R.O.; Nie, X.; Northwood, D.O. A spectroscopic and microstructural study of oxide coatings produced on a Ti-6Al-4V alloy by plasma electrolytic oxidation. *Mater. Chem. Phys.* **2012**, *134*, 484–492. [CrossRef]

15. Liu, R.; Wu, J.; Xue, W.; Qu, Y.; Yang, C.; Wang, B.; Wu, X. Discharge behaviors during plasma electrolytic oxidation on aluminum alloy. *Mater. Chem. Phys.* **2014**, *148*, 284–292. [CrossRef]

16. Dunleavy, C.S.; Golosnoy, I.O.; Curran, J.A.; Clyne, T.W. Characterisation of discharge events during plasma electrolytic oxidation. *Surf. Coat. Technol.* **2009**, *203*, 3410–3419. [CrossRef]

17. Dunleavy, C.S.; Curran, J.A.; Clyne, T.W. Time dependent statistics of plasma discharge parameters during bulk AC plasma electrolytic oxidation of aluminium. *Appl. Surf. Sci.* **2013**, *268*, 397–409. [CrossRef]

18. Parfenov, E.V.; Yerokhin, A.L.; Matthews, A. Frequency response studies for the plasma electrolytic oxidation process. *Surf. Coat. Technol.* **2007**, *201*, 8661–8670. [CrossRef]

19. Yerokhin, A.; Parfenov, E.V.; Liang, C.J.; Mukaeva, V.R.; Matthews, A. System linearity quantification for in-situ impedance spectroscopy of plasma electrolytic oxidation. *Electrochem. Commun.* **2013**, *27*, 137–140. [CrossRef]

20. Boinet, M.; Verdier, S.; Maximovitch, S.; Dalard, F. Plasma electrolytic oxidation of AM60 magnesium alloy: Monitoring by acoustic emission technique. Electrochemical properties of coatings. *Surf. Coat. Technol.* **2005**, *199*, 141–149. [CrossRef]

21. Clyne, T.W.; Troughton, S.C. A review of recent work on discharge characteristics during plasma electrolytic oxidation of various metals. *Int. Mater. Rev.* **2019**, *64*, 127–162. [CrossRef]

22. Sah, S.P.; Tatsuno, Y.; Aoki, Y.; Habazaki, H. Dielectric breakdown and healing of anodic oxide films on aluminum under single pulse anodizing. *Corros. Sci.* **2011**, *53*, 1838–1844. [CrossRef]

23. Aliofkhazraei, M.; Rouhaghdam, A.S. Study of nanoparticle adsorption in single discharge of plasma electrolysis. *Electrochem. Commun.* **2012**, *20*, 88–91. [CrossRef]

24. Zhang, X.; Yao, Z.; Jiang, Z.; Zhang, Y.; Liu, X. Investigation of the plasma electrolytic oxidation of Ti6Al4V under single-pulse power supply. *Corros. Sci.* **2011**, *53*, 2253–2262. [CrossRef]

25. Yang, K.; Cao, B.; Ding, L.; Yang, G.; Huang, Z.H. Design of High Frequency Narrow Pulsed Inverter Power Supply for Micro Arc Oxidation. *J. Sou. Chi. Univer. Technol.* **2014**, *42*, 18–23.

26. Yang, K.; Cao, B. Electrical characteristics identification of dielectric film breakdown during plasma electrolytic oxidation process. *Mater. Lett.* **2015**, *143*, 177–180. [CrossRef]

27. Mécuson, F.J.; Czerwiec, T.; Henrion, G.; Belmonte, T.; Dujardin, L.; Viola, A.; Beauvir, J. Tailored aluminum oxide layers by bipolar current adjustment in the Plasma Electrolytic Oxidation (PEO) process. *Surf. Coat. Technol.* **2007**, *201*, 8677–8682. [CrossRef]

 *materials*

Article

Hot-Probe Characterization of Transparent Conductive Thin Films

Alexander Axelevitch

Engineering Faculty, Holon Institute of Technology (HIT), Holon 5810201, Israel; alex_a@hit.ac.il;
Tel.: +972-544-718-122

Abstract: Transparent conductive oxide (TCO) thin films represent a large class of wide-bandgap semiconductors applied in all fields of micro- and optoelectronics. The most widespread material applied for the creation of TCO coatings is indium-tin oxide (ITO). At the same time, there are plurality trends to change the high-cost ITO on other materials, for example, on the ZnO doped by different elements such as Al, Mn, and Sb. These films require mobile and low-cost evaluation methods. The dynamic hot-probe measurement system is one of such techniques that can supplement and sometimes replace existing heavy systems such as the Hall effect measurements or the Haynes–Shockley experiments. The theoretical basis and the method of analysis of the recorded dynamic hot-probe characteristics measured at different temperatures were presented in this work. This method makes it possible to extract the main parameters of thin films. Commercial thin ITO films and new transparent conducting ZnO:Al layers prepared by magnetron co-sputtering were studied by the proposed method. The measured parameters of commercial ITO films are in agreement with the presented and reference data. In addition, the parameters of ZnO:Al thin films such as the majority charge carriers type, concentration, and mobility were extracted from dynamic hot-probe characteristics. This method may be applied also to other wide-bandgap semiconductors.

Keywords: dynamic hot-probe measurements; indium-tin oxide; aluminum-zinc oxide; magnetron co-sputtering

Citation: Axelevitch, A. Hot-Probe Characterization of Transparent Conductive Thin Films. *Materials* **2021**, *14*, 1186. https://doi.org/10.3390/ma14051186

Academic Editor: Erwan Rauwel

Received: 8 February 2021
Accepted: 26 February 2021
Published: 3 March 2021

Publisher's Note: MDPI stays neutral with regard to jurisdictional claims in published maps and institutional affiliations.

1. Introduction

Transparent conductive thin films play a very important role in microelectronics and optoelectronics nowadays. Thin films such as ITO (indium-tin oxide) have become the standard coating used in all sides of the micro- and optoelectronics. The ITO thin films found applications in solar cells, flat panel displays, flat transparent heaters, electrochromic windows, etc. At the same time, there are many trends to substitute the high-cost indium by other materials. Transparent conductive zinc oxide (ZnO) films are attractive semiconductor coatings that may substitute ITO films in high-temperature applications (higher than 700 K) [1]. Due to the impurity concentration and the thin films structure, these films can have a resistivity from 10^{10} to 10^{-4} $\Omega \cdot cm$ [2]. There are many methods for producing doped ZnO films: thermal evaporation using an electron beam [3], sputtering in a DC plasma [2] and in a RF plasma [4], sputtering from a mosaic target [5], and simultaneous sputtering from two different targets [6]. The method of simultaneous sputtering from two targets was used in this work. This method allows for a fairly accurate control of the alloy metal percentage by controlling the power of the spray sources and the distance between the sputtering sources and the substrate.

All thin-film semiconductor materials require the fine control of their characteristics for successful applications. Transparent conductive oxides are wide-bandgap semiconductors, usually of the degenerate type, that is, having the high concentration of majority carriers. Their properties significantly depend on their growth conditions. Thus, the indium oxide, In_2O_3, has a bandgap of ≈ 3.7 eV [7]. To obtain enough high conductivity while maintaining

117

high transparency, this material is applied with alloyed tin (Sn) in relation 5–10%. Sn produces interstitial bonds with oxygen inside the In_2O_3 structure by replacing the In^{3+} atoms on the Sn^{2+} and Sn^{4+}, which exist either as SnO or SnO_2 states, respectively. However, a predominance of the higher valency in Sn acts as an n-type donor releasing electrons to the conducting band. Thus, in ITO, both interstitial tin and oxygen vacancies contribute to the high conductivity [8].

Zinc oxide, ZnO, is a wide-bandgap semiconductor of class A_2B_6 with a direct bandgap of width ≈ 3.4 eV [9]. To increase the conductivity of ZnO thin films, doping with various material, such as Al, Sn, and other, was used [2,10]. Al-doped ZnO (AZO) is the most widespread transparent conductive coating (TCO) based on the ZnO. Due to the wide bandgap, perfect ZnO crystals have low conductivity due to the limited (low) number of intrinsic charge carriers at room temperature (RT). Therefore, defects begin to play a decisive role in the conductivity of wide-bandgap semiconductors. Thus, two type of defects, substitutional Al^{3+}, replacing Zn^{2+}, acting as donors and oxygen vacancies define electrical properties of AZO thin films. Table 1 represents the basic average values of the electrical parameters of ITO and AZO [8,11,12].

Table 1. Basic parameters of indium-tin oxide (ITO) and Al-doped ZnO (AZO) thin films.

Parameter	ITO	AZO
Energy gap, Eg (eV)	3.7	3.4
Relative permittivity, $\varepsilon_{r,\infty}$	4	3.66
Static dielectric constant, $\varepsilon_{r,0}$	9	8.91
Electron effective mass, m_n^* (m_0)	0.3	0.24
Hole effective mass, m_p^* (m_0)	0.22	≈ 0.7
Typical electron concentration, n_c (cm^{-3})	$\approx 5 \cdot 10^{20}$	$\approx 10^{19}$
Mobility of electrons, μ_n (cm^{2}v^{-1}s^{-1})	26	50

Properties of TCO coatings are influenced significantly by the deposition method because their conductivity is defined by the density and distribution of defects. Basic characteristics of these coatings such as transmittance in the required range and resistivity are defined by the type of semiconductor, bandgap of material, majority charge carrier's concentration, and the mobility and lifetime of the majority carriers. Usually, these parameters may be measured by such conventional methods as the Hall effect measurement and the Haynes–Shockley experiment, which provide the required data. However, these methods are complex and too expensive. For example, the Hall measurements require the samples with very accurate geometry. Equipment for the Haynes–Shockley experiment uses the specific short-time switching systems for minority charge carrier's excitation. Thus, these methods are available in a limited number of laboratories equipped by these high-cost systems only. The main requirements for evaluation methods are simplicity, efficiency, and sufficient accuracy [13]. Therefore, a simpler and faster method for evaluation of the mentioned parameters is required. The low-cost dynamic hot-probe method [13–15] for the measurement, recording, and extraction of the basic parameters of semiconducting TCO thin films is presented in this work. To verify the proposed method, the commercial ITO coatings and competing home-made AZO thin films prepared using the magnetron co-sputtering technique were investigated using the dynamic hot-probe method.

2. Materials and Methods

2.1. Experimental Details

Two types of transparent conductive coatings were investigated: the commercial ITO-coated glass slides with dimensions 25×75 mm^2 and thickness of 1 mm of "Nanocs" and the glass slides of the same dimensions coated by thin ZnO films doped by Al with various concentrations. Sheet resistance and transmittance in the visible range of commercial ITO thin films are dependent on the film thickness. Usually, these films have a thickness of 15–300 nm, the sheet resistance of 5–100 Ω/sq, and the transmittance is over 85% [16]. The optical properties of the films were measured using the spectrophotometer UV-VIS "BioMate 3S". The films' topography was studied using the metallurgical microscope Hirox RH-2000. The surface resistivity was measured by the standard four-point method using the Macor-probe of MDC. Rough approximate estimation of the films' thickness was carried out by weighing the substrates before and after deposition by using the analytical balance ASB-220-C2. In addition, the thickness of the films can be determined on the basis of the interference fringes of the measured transmittance characteristics. This approach was applied to estimate the thickness of a commercial ITO film. The dynamic hot-probe method was used for characterization of the type, concentration, and mobility of the major charge carriers in commercial ITO coatings and grown AZO thin films. Figure 1 represents the principal measurement scheme (a) and the laboratory home-made setup for dynamic hot-probe measurements (b). To provide measurements, we applied the Digital Multimeter 34405A of Agilent and the software "Keysight BenchVue" providing the real-time measurement and the recording of measured signals. The measurement duration was chosen from 30 to 90 s in order to avoid heating of the second (cold) electrode. It is known that the temperature of the cold electrode does not exceed room temperature by more than 20 °C during the specified time period when temperatures of the hot electrode do not exceed 300 °C [14].

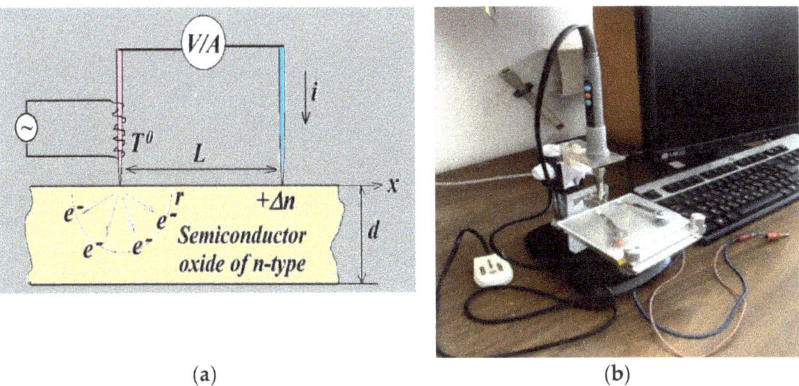

Figure 1. The laboratory setup for dynamic hot-probe measurements: (**a**) the principal operation scheme; (**b**) the laboratory home-made measurement setup.

The principle of operation of this technique is that the heated majority of charge carriers, electrons or holes, have a higher velocity of movement than the cold ones. As soon as we begin to heat one of two electrodes, joining the surface of the semiconductor, the major charge carriers run away from it. As a result, a potential difference arises between the hot and cold electrodes, the sign of which is determined by the type of major charge carriers. If the heated electrode is connected to the positive terminal of the measuring voltmeter, electronic conductivity will show a positive voltage, and a hole-type material will show a negative voltage. To short-circuit the hot and the cold electrodes with an ammeter, a current will begin to flow through it. The sign of the current is defined by the type of the major charge carriers. The bandgap of transparent conductive oxides is large; therefore, the growth in the concentration of intrinsic charge carriers due to the heating of the hot electrode will be vanishingly small. Therefore, only thermal running away of the heated charge carriers should be taken into account through the evaluation process. Analysis of the measured dynamic characteristics of voltage and current dependences on the temperature conditions allows us to determine the recombination rate, the concentration, and the mobility of majority carriers.

ZnO thin films alloyed with Al were deposited simultaneously by the magnetron co-sputtering using a VST service RF-DC sputtering system, model TESF-842, equipped with a turbomolecular pump enabling an ultimate pressure lower than $5 \cdot 10^{-7}$ Torr [17]. The system includes three magnetrons of 2 inches in diameter and enables the simultaneous co-sputtering process using a 13.56 MHz, 300 W RF generator and two DC supplies of 1000 W each. All thin films were deposited at argon atmosphere with the pressure of 3–5 mTorr. Deposition duration was 30 min always. The RF power was supplied to the ZnO target, and the DC power was applied to the Al target. These thin films were grown at room temperature, and they were annealed after deposition in vacuum at 300, 400, and 500 °C through one hour. The amount of Al in the ZnO matrix was controlled by the relation of power applied to the Al target. We tried to grow the ZnO thin films alloyed with approximately 2% weight of aluminum. Evaluation of the thin films composition was provided using Energy Dispersive Spectrometry (EDS) system—Oxford Instruments X-Max^N detector (Czech Republic) completed with the electronic microscope FEG-SEM Tescan Mira 3.

2.2. Theoretical Background

Let us consider once more Figure 1a, the principal operation scheme of the hot-probe measurements. Charge carriers in semiconductors can be in two states: a state of static (thermodynamic) equilibrium and a non-equilibrium state, due to external influences that may be electric, magnetic, or temperature fields. The distribution of charged carriers on energy in the equilibrium state is defined by the Fermi–Dirac equation

$$f_0(E) = \frac{1}{1 + e^{\frac{E - E_F}{k_B T}}},\tag{1}$$

where E_F is a Fermi energy, T is the absolute temperature in Kelvins, and k_B is the Boltzmann constant. At the presence of external influences, the charge carriers system will be described by the non-equilibrium distribution function $f(E) = f(p,r,t)$ dependent on energy (momentum, p), coordinates (r), and time (t). This function, $f(E)$, takes into account all possible mechanisms by which the distribution function may be changed. In real materials, all charge carriers experience collisions with other particles, impurities, phonons, etc. So, the function $f(E)$ should take into account the scattering of charged carriers due to collisions when the function comes to another equilibrium state. If the external influence is finished, the function will return to the initial state. By this way, the distribution function behavior may be presented by three different parts: an excitation, defined by the value of influencing external field; a steady state with slow current processes; and a relaxation part with a return to the initial state.

All these parts may be described by the same general equation taking into account all external and internal processes. In general, the behavior of charged carriers, in particular electrons, can be described by the well-known Boltzmann Transport Equation (BTE), which in differential form looks as follows [18,19]:

$$\frac{\partial f}{\partial t} + \dot{k}\frac{\partial f}{\partial k} + \dot{r}\frac{\partial f}{\partial r} = \left(\frac{\partial f}{\partial t}\right)_{Coll},\tag{2}$$

where the first summand on the left side represents the function variation in time, the second represents the influences of external fields, and the third shows the coordinate variation. The right part of the equation takes into account different types of collisions affecting the motion of particles: scattering by ionized impurities, scattering by neutral impurities, scattering by dislocations, and scattering by grain barriers, which is very important for polycrystalline semiconductor thin films [20]. The right side represents a scattering of the particles and, for simplicity, it may be presented using an electron relaxation time, τ. In the case when there are not external fields and charge carriers, return from the excited state to the initial state occurs as follows (relaxation time approximation):

$$\left(\frac{\partial f}{\partial t}\right)_{Coll} = -\frac{f - f_0}{\tau}. \tag{3}$$

This value, τ, depends on the dominant scattering mechanisms and may be found from experimentally measured dynamic hot-probe characteristics.

To evaluate the recorded dynamic hot-probe characteristics, it is necessary to transform the BTE into an equation without external electrical and magnetic fields; it must take into account only the temperature gradient affecting the movement of charge carriers. Only an internal electric field arises with a directed flow of charge carriers. We also assume that the material being measured is isotropic and the charge carriers flux occurs in only one x direction, where x denotes the r coordinate.

$$F = m\frac{dv}{dt} = \hbar\frac{dk}{dt} = -qE_{ex}, \tag{4}$$

where E_{ex} is an external electrical field. According to Relation (4), the force related term in the full BTE (2) may be described as follows:

$$\dot{k}\frac{\partial f}{\partial k} = -qE_{ex}v\frac{df}{dE_{ex}}, \tag{5}$$

however, in the case of no external electrical field, this term will be vanished.

When the material is heated at a certain point on the surface, the charge carriers begin to move and, thus, they create a current and an electrical field in accordance with the following relationship [21]:

$$E_x = \frac{1}{\sigma}j + \beta\frac{dT}{dx}, \tag{6}$$

where j is the current density, σ is the conductivity of a material dependent on temperature, and β is the additional coefficient characterizing the thermo-electrical properties of the material. This coefficient represents the thermopower (the Seebeck coefficient) produced in the material under non-homogeneous heating due to charged carriers transport [22]. Evidently, this movement occurs only up to reaching of steady state when the current reaches a suitable value j_s (saturation state) and our material will come to the new dynamic steady state (excited state). In this state, the created electrical field is defined by the thermal non-equilibrium and moving charge carriers:

$$E_x = \frac{1}{\sigma}j_s + \beta\frac{dT}{dx}. \tag{7}$$

Equation (7) may be integrated on the distance L, which takes the follows result:

$$U_L = \frac{L}{\sigma}j_s + \beta\Delta T. \tag{8}$$

Here, the conductivity of the sample may be described by the known formulae (the Drude equation):

$$\sigma = \frac{1}{\rho} = q\mu n, \tag{9}$$

where μ is the mobility of charge carriers and n is its concentration. Both parameters, mobility and concentration of charge carriers, are functions of temperature. The mobility of charge carriers, μ, may be found using the well-known Einstein's relation:

$$D = \mu\frac{k_B T_e}{q}, \tag{10}$$

where D is the diffusion coefficient of the major charge carriers, k_B is the Boltzmann's constant, and T_e is an ambient temperature in K. Taking into account that the diffusion distance in our setup is equal to $L = \sqrt{D\tau}$, where τ is the measured relaxation time and substituting with Equation (10), we obtain:

$$\mu = \frac{L^2 q}{k_B T_e \tau}. \tag{11}$$

On the other hand, the mobility of charge carriers characterizes the scattering processes through the average scattering relaxation time, $\langle\tau\rangle$:

$$\mu \cong \frac{q\langle\tau\rangle}{m^*}, \tag{12}$$

where m^* represents an effective mass of electrons. All scattering processes occur simultaneously; therefore, the mobility may be presented by two summands representing the lattice, μ_L, and impurity, μ_i, scattering:

$$\frac{1}{\mu} = \frac{1}{\mu_L} + \frac{1}{\mu_i}. \tag{13}$$

The expressions for these parameters may be presented in following form [23]:

$$\mu_L \propto \frac{4q}{m^\times \sqrt{9\pi k_B}} T^{-1.5}, \tag{14}$$

$$\mu_i \propto \frac{8q k_B^{1.5}}{N_i m^\times \sqrt{\pi}} T^{1.5}, \tag{15}$$

where N_i is the concentration of ionized impurities. Mobility may be found from recorded hot-probe characteristics using a numerical differentiation method, for example, using the known approximate three-point formula of Lagrange [24]:

$$\left\{ \begin{array}{c} f'(t_0) = \frac{1}{2\Delta t}[-3f(t_0) + 4f(t_0 + \Delta t) - f(t_0 + 2\Delta t)] \\ f'(t_0 + \Delta t) = \frac{1}{2\Delta t}[-f(t_0) + f(t_0 + 2\Delta t)] \\ f'(t_0 + 2\Delta t) = \frac{1}{2\Delta t}[f(t_0) - 4f(t_0 + \Delta t) + 3f(t_0 + 2\Delta t)] \end{array} \right. , \tag{16}$$

where a function f represents the measured voltage, t_0 is the first time-point of the decreasing function, and Δt is the time-space between measured points. The thermopower may be found by the same way for suitable temperatures.

Combining Equations (8) and (9), we obtain an equation describing the behavior of the studied material under hot-probe measurement conditions.

$$U_L = \frac{L}{q\mu n} j_s + \beta\Delta T \tag{17}$$

When a heater comes in contact with the electrode, electrons run up into a semi-infinite space of a conductive matter. So, the current density will be related with the measured current, i_s, according to the following equation:

$$j_s = \frac{i_s}{2\pi r^2}, \tag{18}$$

where r is a distance from the heated electrode ($2\pi r^2$ being the surface area of the hemisphere). The propagation of electrons in the space shaped in the form of a thin disc is limited by the disc thickness. Therefore, Equation (18) transforms into the following approximated relation at the distance of integration:

$$j_s = \frac{i_s}{2\pi dL}.$$ (19)

After substitution, Equation (17) transforms into the following final expression:

$$U_L = \frac{i_s}{2\pi q \mu n d} + \beta \Delta T.$$ (20)

This equation, describing the system behavior in both excited and relaxed states, may be solved for the free charge carriers concentration:

$$n = \frac{i_s}{2\pi q \mu d (U_L - \beta \Delta T)}.$$ (21)

Thus, one can calculate the concentration of charge carriers using the experimental data.

The concentration of free charge carriers depends significantly on the thickness of the studied thin film (see Equation (21)). Therefore, the method and accuracy of the thickness measurement will affect the results of our calculations. An error in determining the thickness leads to a deviation in the value of the concentration of charge carriers. Moreover, it is easy to show that the experimental error in determining the concentration of free charge carriers will always be less than the experimental error in measuring the thickness:

$$\frac{\Delta n}{n} = -\frac{\Delta d}{(d + \Delta d)}.$$ (22)

Therefore, the thickness becomes a decisive factor in determining the accuracy of the thin film parameters extracted from the hot-probe characteristics.

A real experiment is a process driven by both variable controlled parameters and random variables that influence the expected results. In the case of dynamic hot-probe measurement, these casual random values can represent a small deviation of the temperature of the heated electrode from the specified one, a change in the contact of this electrode with the sample being measured, etc. Therefore, in order to reduce the influence of various random parameters on the shape of the recorded hot-probe characteristic, it is desirable to carry out several measurements under the same conditions. Recorded characteristics in tabular form should be averaged to use in the following calculations. Obviously, the measurement accuracy increases with the number of measurements. Usually, in order to achieve required process accuracy, it is sufficient to carry out three different measurements at the same temperature and process duration in different places of the studied thin film.

3. Results and Discussion

3.1. Commercial ITO Thin Films

First, we measured the optical properties and sheet resistance of glass slides coated by ITO because the accompanying documents obtained with the commercial slides present the average characteristics only. Figure 2 represents the measured transmittance characteristic of the ITO/glass system.

The inset shows the absolute transmittance characteristic of the ITO films produced by "Nanocs" [16]. The difference between the two curves in Figure 2 can be explained as follows: the manufacturer presents one of the measured or average transmittance characteristics, and the actual characteristics plotted on the graph always have some difference. The sheet resistance of the commercial ITO films measured using the four-point probe method was of $R_\# = 8.3\ \Omega/\text{sq}$, which correlates with the data of the manufacturer (10 Ω/sq).

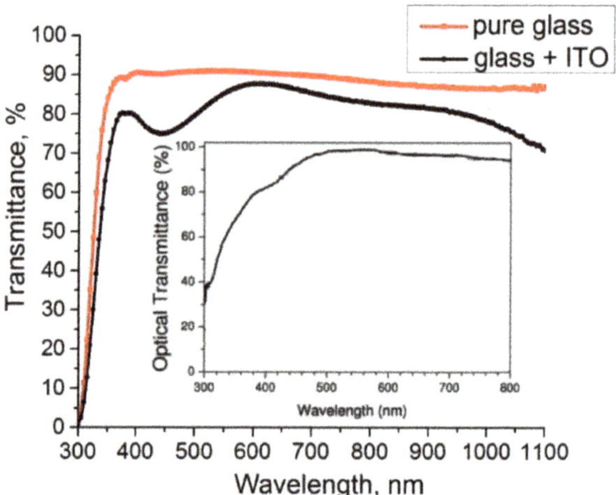

Figure 2. Measured transmittance characteristics of the commercial ITO on glass. The inset represents the absolute transmittance of ITO presented by the manufacturer [16].

To estimate the thickness of the commercial ITO film on the glass slide, the measured transmittance characteristics of this film (see Figure 2) should be considered. The measured spectrum demonstrates periodic oscillations of transmittance due to interference effects in the thin film. These effects are caused by multiple reflections of light from interfaces between the thin film, glass substrate, and air. The transmittance of this system may be described using the following equation [25]:

$$T = \frac{\left[1 - \left(\frac{1-n}{1+n}\right)^2\right]\left[1 - \left(\frac{n-n_{glass}}{n+n_{glass}}\right)^2\right]}{1 + \left(\frac{1-n}{1+n}\right)^2\left(\frac{n-n_{glass}}{n+n_{glass}}\right)^2 - 2\left(\frac{1-n}{1+n}\right)\left(\frac{n-n_{glass}}{n+n_{glass}}\right)cos\left(\frac{4\pi}{\lambda}nd\right)}, \tag{23}$$

where n_{glass} is the refractive index of the glass substrate, $n(\lambda)$ is the refractive index of the film dependent on the light wavelength, λ, and d is the thickness. The maximum and minimum transmittance, according to Formula (23), are defined by the following relations: $2nd = m\lambda$ (m = 1, 2, 3 ...) and $2nd = (2m + 1)\lambda/2$ (m = 0, 1, 2 ...) respectively. Then, one can choose the maximum or minimum points in the measured characteristics and calculate the thickness by choosing and changing the refraction coefficient. Verification should be made by calculation of the transmittance and comparing with the measured values. The calculated thickness of the ITO films was 150 ± 10 nm or ≈150 nm. Then, the resistivity of these films at room temperature can be estimated:

$$\rho = R_\#{\cdot}d = 0.83{\cdot}1.5{\cdot}10^{-4} = 1.25{\cdot}10^{-4} \; [\Omega{\cdot}cm]. \tag{24}$$

Figure 3 represents the dynamic hot-probe characteristics recorded on the commercial glass slides coated by the ITO.

Figure 3a illustrates the dependence of a hot-electrode voltage on the processing time for different temperatures, and Figure 3b shows the current measured between hot and cold electrodes, which were measured for the same temperatures. The curves shown in Figure 3a represent a positive voltage measured between hot and cold electrodes for different temperatures. This shows that the ITO coating is an n-type semiconductor. As shown, the dynamic curves repeat each other; therefore, all dynamic processes in the film are the same for different temperatures. In the recorded plots, three different areas can be identified: the region of a steep increase in voltage due to heating, the region of steady-state voltage, and the region of a sharp decrease in voltage when the heater is disconnected from the hot electrode. All these processes occur in the absence of some external electrical or magnetic fields, so they are driven by the temperature difference only. The steep rise of the voltage between electrodes and the current reflects a run up of hot electrons from the heated electrode. This rise happens very quickly due to the difference between rates of the heated and cold electrons. Evidently, these rates are proportional to the root square of the heating temperature. Therefore, a higher charge difference will be obtained for higher temperature, which is confirmed by measurements. After measurements at elevated temperatures and after the relaxation time, as shown in Figure 3, the properties of films return to their original state.

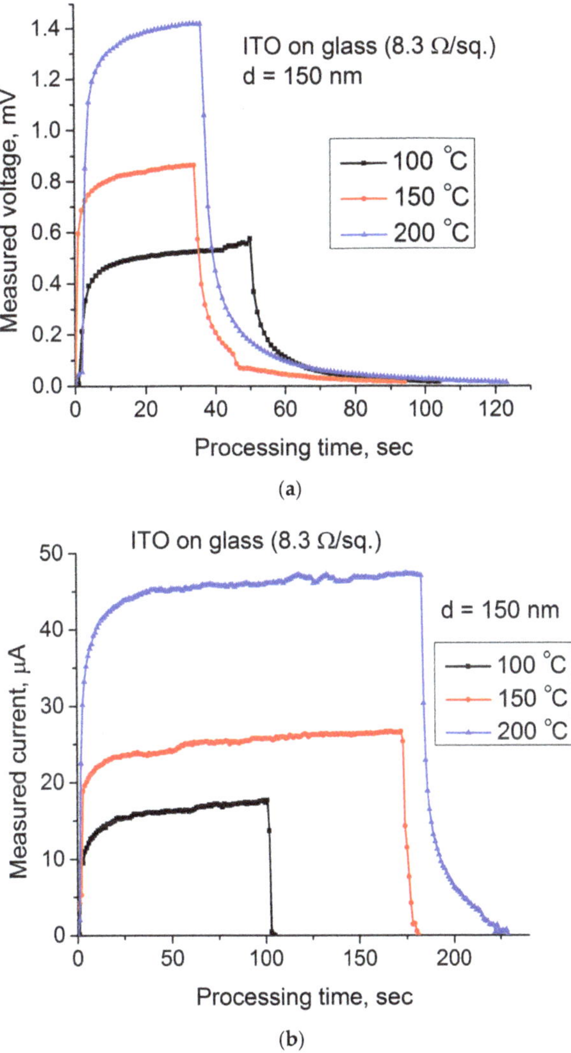

Figure 3. Dynamic hot-probe characteristics of the commercial ITO coatings: (**a**) a voltage measured between hot and cold electrodes; (**b**) a current between electrodes.

3.2. AZO Thin Films Prepared by Co-Sputtering

Thin films ZnO:Al were produced by the co-sputtering method using the simultaneous sputtering of two targets, pure ZnO and pure Al, in the argon atmosphere. To prevent the sputtered particles, including oxygen atoms from the ZnO target, from influencing the Al target and oxidizing it, they were separated by a metal screen. The relation between components was controlled by the variation in their sputtering rate. We prepared the films with 2 and 4 atomic percentage of Al in the composition. Figure 4 illustrates the EDS diagram for the ZnO:Al film containing $\approx$4 atomic percentage of Al. We applied RF power of 200 W for sputtering the ZnO target and DC power of 15 W for the Al target. The estimated thickness of the film was of 860 nm. After deposition, the film was annealed in vacuum conditions ($\approx 5 \cdot 10^{-2}$ Torr) through one hour at temperature of 500 °C.

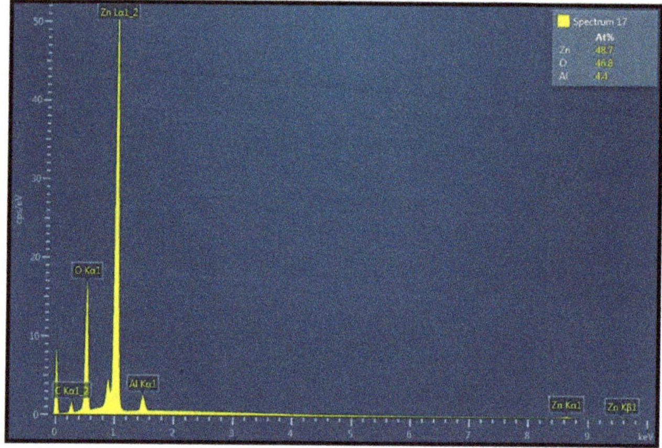

Figure 4. The EDS diagram characterizing a composition of the AZO thin film deposited by the co-sputtering method: the estimated thickness of this film was of $\approx$860 nm; after deposition, the film was annealed at temperature 500 °C through one hour at for-vacuum conditions ($\approx 5 \cdot 10^{-2}$ Torr).

Figure 5 represents transmittance characteristics of two films with different resistivity, low and high.

The difference between two curves was the result of different methods of after-deposition thermal treatment. The high-resistivity film was annealed at air atmosphere pumped up to 0.05 Torr at 500 °C during 1 h. The low-resistivity film was annealed in vacuum conditions of $2 \cdot 10^{-6}$ Torr during the same time and at the same temperature. As shown in Figure 5, the film annealed at high pressure (0.05 Torr) is more transparent than that explained by good oxidation. The difference in the partial pressure of oxygen in both heat treatment procedures was approximately four orders of magnitude. Thus, oxidation takes place entirely at principal higher pressure, which leads to an increase in transmittance and resistivity of the resulting thin films.

Figure 6 represents the dynamic hot-probe characteristics measured on the high-resistive ZnO:Al coating deposited on the usual glass slide.

Figure 6a illustrates the dependence of a hot-electrode voltage on the processing time for different temperatures, and Figure 6b shows the current measured between hot and cold electrodes recorded for the same temperatures. These characteristics are noisy; moreover, the current was measured near the limit of sensitivity of the measuring device. Apparently, the observed oscillations are associated with the high resistivity of measured samples. Obviously, this claim requires statistical proof. This will be the focus of future studies. Figure 7 represents the dynamic hot-probe characteristics measured on the low-resistive ZnO:Al coating deposited on the usual glass slide.

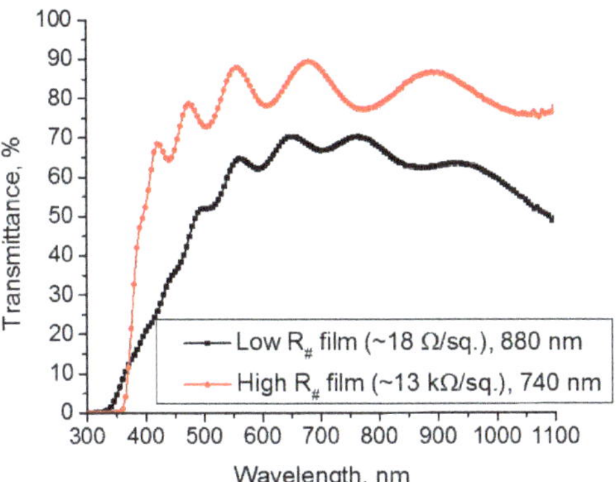

Figure 5. Transmittance characteristics of AZO thin films with low and high resistivity.

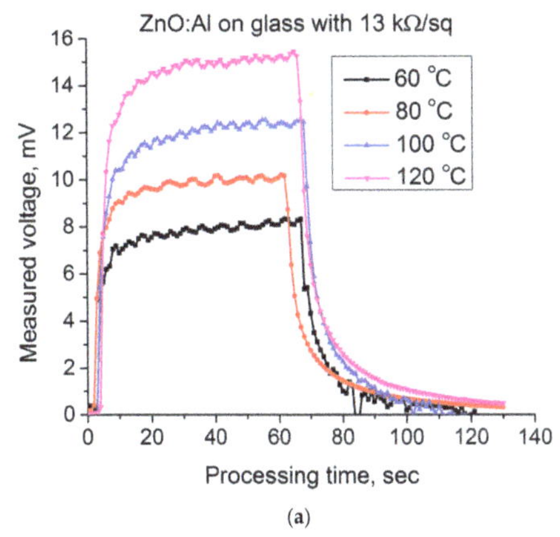

(a)

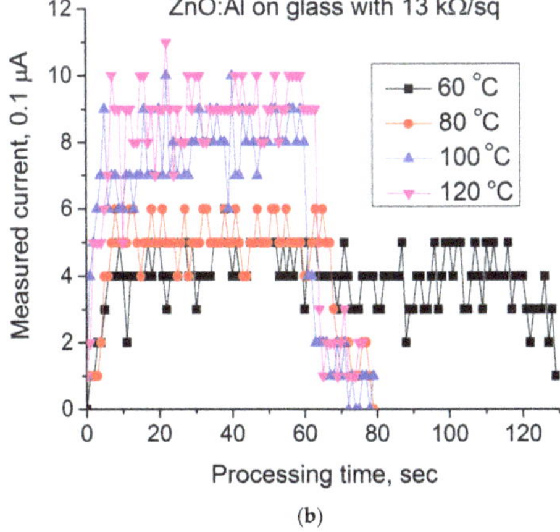

(b)

Figure 6. Dynamic hot-probe characteristics of the high-resistive AZO coatings: (a) a voltage measured between hot and cold electrodes; (b) a current between electrodes.

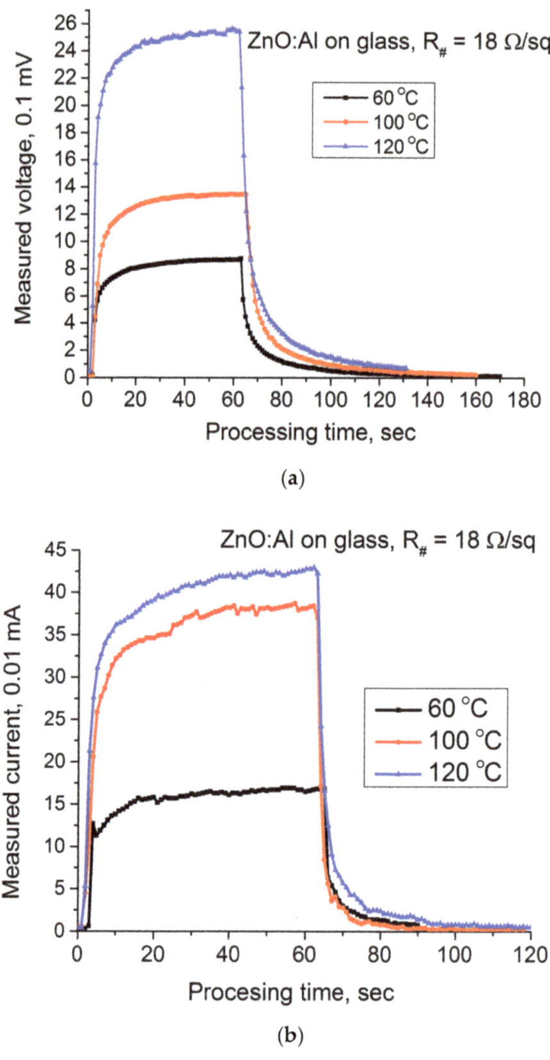

(**a**)

(**b**)

Figure 7. Dynamic hot-probe characteristics of the low-resistive AZO coatings: (**a**) a voltage measured between hot and cold electrodes; (**b**) a current between electrodes.

Figure 7a illustrates the dependence of a hot-electrode voltage on the processing time for different temperatures, and Figure 7b presents the current measured between hot and cold electrodes measured for the same temperatures. As it was mentioned above, the difference in these figures is due to the different sheet resistance caused by the different oxidation conditions.

3.3. Processing the Experimental Results

Let us consider again Figure 3a recorded for commercial ITO thin films. The measured values of the voltage decreasing after removing a heater from the hot electrode are presented in Table 2. This table also contains a calculation of several parameters, which was performed using the measured data. By definition, the relaxation time may be calculated using a derivative in the first point of the diagram (see Equation (3)):

$$f(t) = f'(t_0)\tau + f(t_i); \quad \tau = -\frac{f(t) - f(t_i)}{f'(t_0)}. \tag{25}$$

Table 2. Measured voltage and basic calculated parameters of the ITO film.

Parameter	Equation	T_0 = 373 K (100 °C)	T_1 = 423 K (150 °C)	T_2 = 473 K (200 °C)
$V(t_0)$		0.576	0.865	1.420
$V(t_1)$		0.368	0.573	1.060
$V(t_2)$		0.288	0.397	0.703
y_0'	16	−0.272	−0.349	−0.362
τ(s)	25	2.12	2.48	3.92
μ (cm^2/V·s)	11	14.74	10.90	6.22
β (μV/K)	16	3.12	8.44	13.76

Here, $f(t_i)$ is the initial value of the measured parameter, a voltage.

Now, using the three-point Formula (16) and the Relation (22), we can calculate the relaxation time for all three diagrams, as shown in Figure 8.

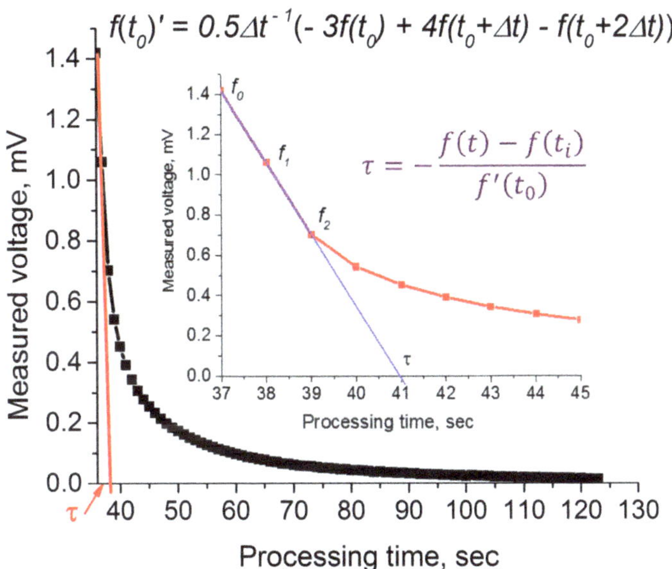

Figure 8. Derivation of the relaxation time from hot-probe measurement; inset shows the characteristic processing.

The thermopower of the ITO film can be calculated using Figure 9, which shows the dependence of the potential difference on the temperature of the heated electrode.

The results obtained for the mobilities show that they obey Formula (14). So, the main relaxation process in industrial ITO films is lattice scattering or the interaction of excited electrons with phonons.

The calculation of the number of charge carriers moving under thermal excitation can be performed using Formula (21) using the first (ascending) part of the hot-probe characteristic. Figure 10 illustrates the processing of the first part of the recorded hot-probe characteristics.

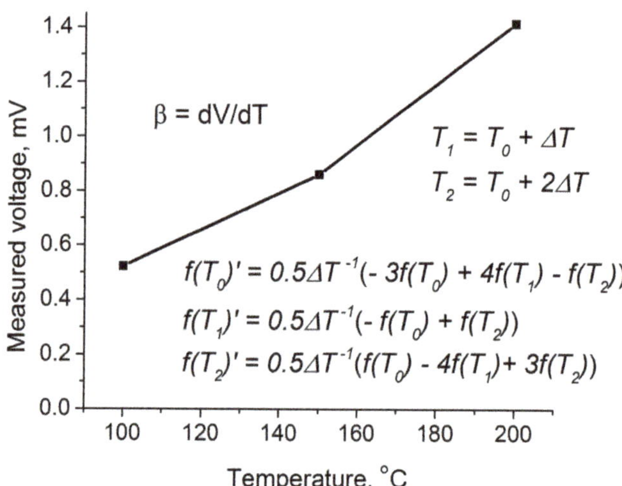

Figure 9. Dependence of the potential difference on the temperature.

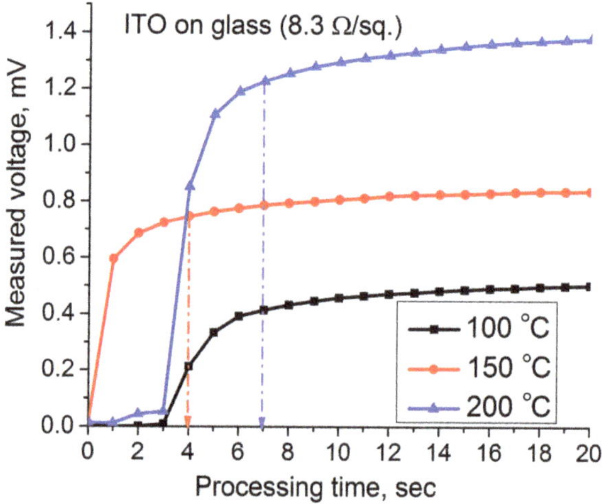

Figure 10. Extraction of saturation parameters, voltage and current, from recorded dynamic hot-probe characteristics.

For the calculation, we use the steady-state values of the measured voltage and current reached within a certain short time. These parameters, together with the calculated mobility and thermopower, make it possible to calculate the concentration of charge carriers and the conductivity of the coating by Formula (9). The calculation results are presented in Table 3:

Comparison of the calculated resistivity of ITO films provided using Formula (9) with the measured value (see Equation (24)) shows good coincidence. Using the calculated conductivity from Tables 4 and 5 and Arrhenius's relation [23], we can estimate the bandgap of the ITO layer, as shown in Figure 11:

$$f(t) = \sigma_T = \sigma_0 e^{\frac{E_g}{2k_B T}}; \quad E_g = 2k_B \frac{ln\sigma_1 - ln\sigma_3}{T_3^{-1} - T_1^{-1}}. \tag{26}$$

Table 3. Measured voltage and basic calculated parameters of the ITO film.

Temperature, T (°C)	Rise Time, Δt (s)	Measured Saturation Current, i_s (µA)	Measured Saturation voltage, U_L (mV)	$n \cdot 10^{20}$ (cm^{-3}) Equation (21)	σ ($\Omega \cdot$cm)$^{-1}$ Equation (9)
100	4	11.5	0.414	2.64	$0.63 \cdot 10^3$
150	4	19.6	0.747	3.67	$0.64 \cdot 10^3$
200	4	35.2	1.227	6.93	$0.69 \cdot 10^3$

Table 4. Measured voltage and basic calculated parameters of the high-resistance AZO film.

Parameter	Equation	$T_0 = 333$ K (60 °C)	$T_1 = 353$ K (80 °C)	$T_2 = 373$ K (100 °C)
$V(t_0)$		0.832	1.017	1.246
$V(t_1)$		0.542	0.872	0.973
$V(t_2)$		0.432	0.638	0.735
$y_0{}'$	16	−0.38	−0.1	−0.29
τ(s)	25	2.19	10.2	4.3
μ (cm^2/V·s)	11	15.75	3.16	7.27
β (µV/K)	16	8.15	10.4	12.6

Table 5. Measured voltage and basic calculated parameters of the high-resistance AZO film.

Temperature, T (°C)	Rise Time, Δt (s)	Measured Saturation Current, i_s (µA)	Measured Saturation Voltage, U_L (mV)	$n \cdot 10^{18}$ (cm^{-3}) Equation (21)	σ ($\Omega \cdot$cm)$^{-1}$ Equation (9)
60	6	0.4	0.609	0.76	1.92
80	6	0.5	0.838	3.38	1.71
100	6	0.7	1.037	1.65	1.92

The bandgap of the commercial ITO film calculated by Formula (26) is equal to 3.38 eV.

Now, one can calculate the parameters of AZO experimental samples, which showed two different resistivities: low and high. Table 4, which is similar to Table 2, represents the measured data and calculated parameters for the AZO thin film with high sheet resistance.

The calculation of the concentration of charge carriers and conductivity obtained under thermal activation can be provided by using the table similar to Table 3.

The calculation of parameters of AZO experimental samples with low sheet resistance can be provided by the same scheme. Measured and calculated parameters are given in Table 6.

The calculation of the charge carriers' concentration and the conductivity obtained under thermal activation is presented in Table 7.

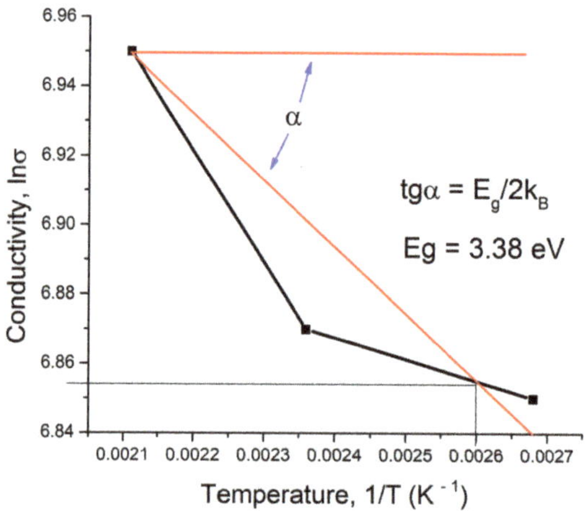

Figure 11. Evaluation of the semiconductor bandgap by experimentally obtained conductivity.

Table 6. Measured voltage and basic calculated parameters of the low-resistance AZO film.

Parameter	Equation	$T_0 = 333$ K (60 °C)	$T_1 = 373$ K (100 °C)	$T_2 = 413$ K (140 °C)
$V(t_0)$		0.876	1.341	2.544
$V(t_1)$		0.575	1.099	2.132
$V(t_2)$		0.447	0.882	1.634
y_0'	A18	−0.388	−0.255	−0.369
τ(s)	A25	2.26	5.26	6.89
μ (cm^2/V·s)	A13	15.26	5.94	4.03
β (μV/K)	A18	3.2	27.8	52.4

Table 7. Measured voltage and basic calculated parameters of the low-resistance AZO film.

Temperature, T (°C)	Rise Time, Δt (s)	Measured Saturation Current, i_s (μA)	Measured Saturation Voltage, U_L (mV)	$n \cdot 10^{18}$ (cm^{-3}) Equation (21)	σ ($\Omega \cdot$cm)$^{-1}$ Equation (9)
60	6	13.1	0.681	0.17	41.5
100	6	17.1	1.055	0.15	14.3
140	6	28.7	2.133	1.43	92.2

4. Conclusions

The paper presents a method for extracting the main parameters of wide-bandgap TCO thin films from the measured and recorded dynamic hot-probe characteristics and its theoretical foundations. The measurement technique is simple and inexpensive. Conventional measuring instruments equipped with simple software allow the recording and storage of measured data. The technique for extracting the main parameters has been tested on commercial ITO thin films. The results are consistent with the presented and reference data. Then, this method was used to study homemade thin films of AZO, which were prepared by the magnetron co-sputtering method. The parameters of these films such as the majority charge carrier type, concentration, and mobility were extracted from the dynamic hot-probe characteristics measured at different temperatures. The results obtained are in agreement with the literature data for thin AZO films. In addition, the proposed method can be used to calculate the bandgap and thermoelectric power of the films.

Funding: This research received no external funding.

Institutional Review Board Statement: No applicable.

Informed Consent Statement: Not applicable.

Data Availability Statement: Not applicable.

Acknowledgments: The authors acknowledge the Holon Institute of Technology for the possibility to provide this research.

Conflicts of Interest: The authors declare no conflict of interest.

References

1. Qadri, S.B.; Kim, H.; Khan, H.R.; Piqué, A.; Horwitz, J.S.; Chrisey, D.; Kim, W.J.; Skelton, E.F. Transparent conducting films of In2O3–ZrO2, SnO2–ZrO2 and ZnO–ZrO2. *Thin Solid Films* **2000**, *377*, 750–754. [CrossRef]
2. Suchea, M.; Christoulakis, S.; Katsarakis, N.; Kiriakidis, G. Comparative study of zinc oxide and aluminum doped zinc oxide transparent thin films grown by direct current magnetron sputtering. *Thin Solid Films* **2007**, *515*, 6562–6566. [CrossRef]
3. Varnamkhasti, M.G.; Fallah, H.R.; Zadsar, M. Effect of heat treatment on characteristics of nanocrystalline ZnO films by electron beam evaporation. *Vacuum 2012 86*, 871–875. [CrossRef]
4. Abdallah, B.; Jazmatia, A.K.; Refaai, R. Oxygen Effect on Structural and Optical Properties of ZnO Thin Films Deposited by RF Magnetron Sputtering. *Mat. Res.* **2017**, *20*, 607–612. [CrossRef]
5. Becerril, M.; Silva-Lopez, H.; Guillen-Cervantes, A.; Zelaya-Angel, O. Aluminum-doped ZnO polycrystalline films prepared by co-sputtering of a ZnO-Al target. *Rev. Mex. Fis.* **2014**, *60*, 27–31.
6. Chen, H.; Qiu, C.; Peng, H.; Xie, Z.; Wong, M.; Kwok, H.S. Co-sputtered Aluminum Doped Zinc Oxide Thin Film as Transparent Anode for Organic Light-emitting Diodes. In Proceedings of the 8th Asian Symposium on Information Display ASID'04, Nanjing, China, 15–17 February 2004; pp. 489–492.
7. Kiriakidis, G.G.; Ouacha, H.; Katsarakis, N. InO$_X$ nanostructured thin films: Electrical and sensing characterization. *Rev. Adv. Mater. Sci.* **2003**, *4*, 32–40.
8. Bashar, S.A. Study of Indium Tin Oxide (ITO) for Novel Optoelectronic Devices. Ph.D. Thesis, University of London, London, UK, 1998.
9. Ellmer, K.; Klein, A.; Rech, B. *Transparent Conductive Zinc Oxide*; Springer: Berlin, Germany, 2007.
10. Alsmadi, A.M.; Masmali, N.; Jia, H.; Guenther, J.; Abu Jeib, H.; Kerr, L.L.; Eid, K.F. Hot Probe Measurements of n-Type Conduction in Sb-doped ZnO Microwires. *J. Appl. Phys.* **2015**, *117*, 155703. [CrossRef]
11. Preissler, N.; Bierwagen, O.; Ramu, A.T.; Speck, J.S. Electrical transport, electrothermal transport, and effective electron mass in single-crystalline In$_2$O$_3$ films. *Phys. Rev. B* **2013**, *88*, 085305. [CrossRef]
12. Quemener, V. Electrical Characterization of Bulk and Thin Film Zinc Oxide. Ph.D. Thesis, University of Oslo, Oslo, Norway, 2012.
13. Axelevitch, A.; Golan, G. Hot-probe method for evaluation of majority charged carriers concentration in semiconductor thin films. *Facta Univ. Ser. Electron. Energ.* **2013**, *26*, 187–195. [CrossRef]
14. Golan, G.; Axelevitch, A.; Gorenstein, B.; Manevich, V. Hot-Probe Method for Evaluation of Impurities Concentration in Semiconductors. *Microelectron. J.* **2006**, *37*, 910–915. [CrossRef]
15. Akter, N.; Afrin, S.; Hossion, A.; Kabir, A.; Akter, S.; Mahmood, Z.H. Evaluation of Majority Charge Carrier and Impurity Concentration Using Hot Probe Method for Mono Crystalline Silicon (100) Wafer. *Int. J. Adv. Mater. Sci. Eng. (IJAMSE)* **2015**, *4*, 13–21. [CrossRef]
16. Nanocs Inc. 2018. Available online: http://www.nanocs.net (accessed on 3 March 2021).
17. Briot, O.; Moret, M.; Barbier, C.; Tiberj, A.; Peyre, H.; Sagna, A.; Contreras, S. Optimization of the properties of the molybdenum back contact deposited by radiofrequency sputtering for Cu(In$_{1-x}$Ga$_x$)Se$_2$ solar cells. *Sol. Energ. Mat. Sol. C* **2018**, *174*, 418–422. [CrossRef]
18. *Lee, HoSung Thermoelectrics: Design and Materials*; John Wiley & Sons: Hoboken, NJ, USA, 2017.
19. Geng, H.; Deng, W.-Y.; Ren, Y.-J.; Sheng, L.; Xing, D.-Y. Unified semiclassical approach to electronic transport from diffusive to ballistic regimes. *Chin. Phys. B* **2016**, *25*, 097201. [CrossRef]
20. Ellmer, K.; Mientus, R. Carrier transport in polycrystalline transparent conductive oxides: A comparative study of zinc oxide and indium oxide. *Thin Solid Films* **2008**, *516*, 4620–4627. [CrossRef]
21. Landau, L.D.; Lifschitz, E.M. *Electrodynamics of Continuous Media*; Science: Moscow, Russia, 1982. (In Russian)
22. Kasap, S.O. *Principles of Electrical Engineering Materials and Design*; McGraw-Hill: Boston, MA, USA, 1997.
23. Kasap, S.; Koughia, C.; Ruda, H.E. Electrical Conduction in Metals and Semiconductors. In *Springer Handbook of Electronic and Photonic Materials*, 2nd ed.; Kasap, S., Capper, P., Eds.; Springer: Cham, Switzerland, 2017.
24. Burden, A.M.; Burden, R.L.; Faires, J.D. *Numerical Analysis*, 10th ed.; Cengage: South Melbourne, Australia, 2016.
25. Axelevitch, A.; Apter, B. Preparation and study of doped ZnS thin films. *Microelelectron. Eng.* **2017**, *170*, 39–43. [CrossRef]

MDPI
St. Alban-Anlage 66
4052 Basel
Switzerland
Tel. +41 61 683 77 34
Fax +41 61 302 89 18
www.mdpi.com

Materials Editorial Office
E-mail: materials@mdpi.com
www.mdpi.com/journal/materials